International Symposium on Reactor Dosimetry (ISRD 17)

EPJ Web of Conferences Volume 278 (2023)

Lausanne, Switzerland
21 – 26 May 2023

Editors:

Pavel Frajtag
Dean Thornton
Jan Wagemans

ISBN: 978-1-7138-7052-4

Printed from e-media with permission by:

Curran Associates, Inc.
57 Morehouse Lane
Red Hook, NY 12571

Some format issues inherent in the e-media version may also appear in this print version.

For additional information, please contact EDP Sciences – Web of Conferences
at the address below.

EDP Sciences – Web of Conferences
17, Avenue du Hoggar
Parc d'Activité de Courtabœuf
BP 112
F-91944 Les Ulis Cedex A
France

Phone: +33 (0) 1 69 18 75 75
Fax: +33 (0) 1 69 28 84 91

contact-edps@webofconferences.org

Additional copies of this publication are available from:

Curran Associates, Inc.
57 Morehouse Lane
Red Hook, NY 12571 USA
Phone: 845-758-0400
Fax: 845-758-2633
Email: curran@proceedings.com
Web: www.proceedings.com

TABLE OF CONTENTS

edp sciences

EPJ Web of Conferences

Statement of Peer review

In submitting conference proceedings to EPJ Web of Conferences, I certify to the Publisher that I adhere to the **Policy on Publishing Integrity** of the journal in order to safeguard good scientific practice in publishing.

1. All articles have been subjected to peer review administered by the proceedings editors.
2. Reviews have been conducted by expert referees, who have been requested to provide unbiased and constructive comments aimed, whenever possible, at improving the work.
3. Proceedings editors have taken all reasonable steps to ensure the quality of the materials they publish and their decision to accept or reject a paper for publication has been based only on the merits of the work and the relevance to the journal.

Title, date and place of the conference

Seventeenth International Symposium on Reactor Dosimetry
21-26 May 2023
École Polytechnique Fédérale de Lausanne (EPFL), Lausanne, Switzerland.

Proceedings editor(s):

Pavel Frajtag
Dean Thornton
Jan Wagemans

(in alphabetical order)

Date and editor's signature

10th February 2023

Fluorescence correction for activity measurement of ^{93m}Nb in niobium dosimeters: calculation and experimental validation

Marie-Christine Lépy[1*], *Jonathan* Riffaud[1], *Christophe* Domergue[2], *Hervé* Philibert[2], *Christophe* Destouches[2], and *Hubert* Carcreff[3]

[1]Université Paris-Saclay, CEA, LIST, Laboratoire National Henri Becquerel (LNE-LNHB), 91120 Palaiseau, France

[2]CEA, DES/IRESNE/DER/SPESI, Laboratoire de Dosimétrie, Capteurs et Instrumentation (LDCI), Cadarache, 13108 Saint Paul-lez-Durance, France

[3]Université Paris-Saclay, CEA, ISAS/DM2S, Service d'études des réacteurs et de mathématiques appliquées, 91191 Gif-sur-Yvette cedex, France

Abstract. Niobium is a dosimeter used to characterize fast-neutron reaction rates to monitor nuclear reactor vessels. Its characterization is based on the activity of ^{93m}Nb resulting from the ^{93}Nb(n,n') activation reaction. The decay of ^{93m}Nb results mainly in the emission of niobium K X-rays which are used to determine the activity of ^{93m}Nb. Direct measurement using X-ray spectrometry does not require any sample preparation, but fluorescence effects, due to impurities which are activated during irradiation, must be taken into account. Indeed, some of these radioactive impurities, in particular ^{182}Ta, and other niobium isotopes remain present during the measurements and disturb the X-ray spectra. The fluorescence effect induces additional niobium X-ray emission to that due to the ^{93m}Nb decay alone, which leads to an overestimation of the dosimeter activity, especially if the results are expected shortly after the end of the irradiation. It is therefore necessary to assess the contribution of fluorescence effects to provide accurate values of ^{93m}Nb activity. Fluorescence correction factors were established by Monte Carlo simulation. The calculated fluorescence correction factors were validated by an experimental approach, using activated niobium dosimeters with different tantalum concentrations.

1 Introduction

Niobium is a dosimeter used to characterize fast-neutron reaction rates to monitor neutron damage to nuclear reactor vessels [1]. Its characterization is based on the activity of ^{93m}Nb resulting from the activation reaction ^{93}Nb(n, n'). ^{93m}Nb is a metastable state of ^{93}Nb that de-excites towards the ground state by an isomeric transition with a strong internal conversion coefficient. This decay results mainly in emission of niobium K X-rays which are used to determine the activity of ^{93m}Nb, either by direct measurement using X-ray spectrometry or,

[*] Corresponding author : marie-christine.lepy@cea.fr

after sample dissolution, using X-ray counting or liquid scintillation measurement. The first method does not require any sample preparation, but fluorescence effects can disturb the measurement, since niobium foils may contain impurities that are activated during irradiation. Some of these radioactive impurities, especially ^{182}Ta, and other isotopes of niobium (^{94}Nb, ^{95}Nb and ^{92m}Nb) are still present during spectrometry measurements and disturb the X-ray spectra. The fluorescence effect generates additional niobium X-rays to those due only to the decay of ^{93m}Nb, thus inducing an overestimation of the dosimeter activity. Most of the activated impurities have a much shorter half-life than that of ^{93m}Nb and the current method consists in letting the dosimeters decay for several months, or even years if one considers the half-life of ^{182}Ta (114 days), before proceeding with the measurement to provide a relevant activity. However, in some circumstances, results are expected much earlier, no long after the end of irradiation. It is therefore necessary to take into account the effects of fluorescence to provide accurate values of the activity of ^{93m}Nb. The present study focuses on fluorescence corrective factors with a two-steps approach. First, the correction factors were established by Monte Carlo simulation. Second, these were validated by experimental measurements of niobium dosimeters with different tantalum levels that were submitted to an intense neutron flux. These dosimeters were measured six times throughout the year following the end of irradiation using gamma- and X-ray spectrometry to determine respectively the activity of the impurities and the niobium K X-rays emission rates. These rates were corrected for the impurities contribution using the calculated corrective factors before deriving the activity of ^{93m}Nb at the same reference date.

2 Niobium dosimeters preparation and measurement

2.1 Niobium activation

Different reactions occur during neutron activation of niobium dosimeters and two kinds of radionuclides are produced. These result directly from the activation of niobium (Fig. 1) since, in addition to the reaction of interest (n, n'), other processes such as (n, 2n), (n, γ) or (n, p) can provide ^{92}Nb and ^{92m}Nb, ^{94}Nb and ^{93}Zr, respectively. Moreover, in its natural state, niobium is usually found mixed with various other metals, notably tantalum. These are also activated and produce additional radionuclides, e.g. ^{182}Ta is produced by radiative capture from ^{181}Ta. Table 1 shows the main characteristics of the gamma-emitting impurities.

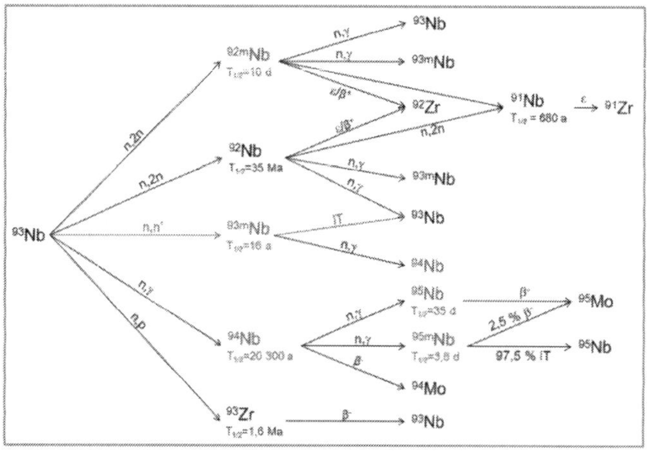

Fig. 1. Activation products from niobium-93

As the presence of tantalum leads to significant corrective factors, for this exercise it was chosen to work with different levels of tantalum impurity concentration. Nine niobium dosimeters with different Ta impurity levels, namely 0.3 ppm, 4 ppm and 19.6 ppm were activated in the MARIA reactor (National Centre for Nuclear Research, Poland) within the framework of the round robin initiated by the European Working Group on Reactor Dosimetry (EWGRD) [2]. The dosimeters were available as strips or discs, with the following characteristics: strips: thickness 20 μm, length 5 mm, width 1 mm, mass 1 mg; discs: thickness 20 μm, diameter 8 mm, mass 8.6 mg.

Table 1. Main characteristics of ^{93m}Nb and of gamma-emitting impurities

Radionuclide	Half-life	Energy (keV)	Emission intensity (%)	Reference
^{93m}Nb	16.12 (15) a	16.58 18.67	9.66 (17) 1.89 (4)	DDEP – KRI 2013 [4]
^{92m}Nb	10.15 (2) d	934.44	99.07 (4)	NDS 66 – 1999 [3]
^{94}Nb	20.0 (24) 10^3 a	702.62 871.09	97.9 (20) 99.9 (1)	NDS 44 & 66 - 1999[3]
^{95}Nb	34.991 (6) d	765.80	99.808 (7)	DDEP – INEL 2002[4]
^{95m}Nb	3.61 (3) d	235.69	25.1 (3)	DDEP – LPRI-INEL 202[4]
^{182}Ta	114.61 (13) d	1 121.29 1 221.39	35.17 (33) 27.27 (27)	DDEP – LNHB 2011[4]
^{183}Ta	5.1 (1) d	246.06	27.2	NDS [3]

2.2 Gamma spectrometry

Gamma spectrometry was used to quantify the activity of radioactive impurities. The spectrometer is based on an N-type high-purity germanium (HPGe) detector whose full-energy peak efficiency has been accurately determined using radioactive standards. A spectrum example is shown in Fig. 2 where the regions corresponding to impurities appear in red. The activity calculation is derived from the full-energy peaks corresponding to the characteristic energies shown in Table 1.

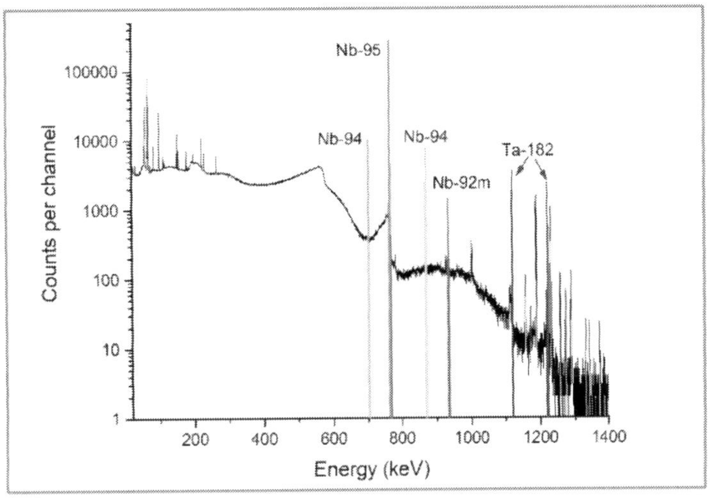

Fig. 2. Gamma spectrum in the 15 keV-1500 keV region allowing the quantification of impurities

The activity, A (Bq), of each radionuclide is calculated from the net area of the full-energy peaks corresponding to the energies of interest, $N(E)$, the acquisition time, t (s) (active time), the associated photon emission intensity, $I(E)$ (see Table 1), the full-energy peak efficiency for the energy E, $\varepsilon(E)$, and various corrective factors, C_i. These include radioactive decay, coincidence summing and self-attenuation [5].

$$A = \frac{N(E) \cdot \prod C_i}{\varepsilon(E) \cdot I(E) \cdot t} \tag{1}$$

2.3 X-ray spectrometry

X-ray spectrometry measurements were carried out with a spectrometer built around an N-type HPGe detector, adapted to low energies and equipped with two collimators to reduce scattering effects. As for the gamma-ray spectrometer, the full-energy peak efficiency has been accurately determined using radioactive standards. Fig. 3 illustrates the detailed processing of the region of interest which includes the peaks corresponding to K X-rays of niobium (16.6 keV and 18.7 keV) and zirconium (15.8 keV and 17.7 keV) as well as the Kα peak of molybdenum (17.4 keV). The peak areas are obtained using the COLEGRAM software [6] by fitting peak functions to the experimental points. Each peak is represented by a Voigt function, i.e. the convolution of a Lorentzian, which allows for the natural width of the X-ray lines, with a Gaussian, which reproduces the broadening of the detection process. The activity of each dosimeter is calculated with Eq. 1 for each niobium peak and the final result is the weighted mean of the two values. Due to the low energies considered, the self-attenuation correction factors, of an order of magnitude of 10 %, were computed by using the software ETNA [7]. In addition, corrective factors had to be calculated to take into account additional fluorescence induced by impurities (see §2.4).

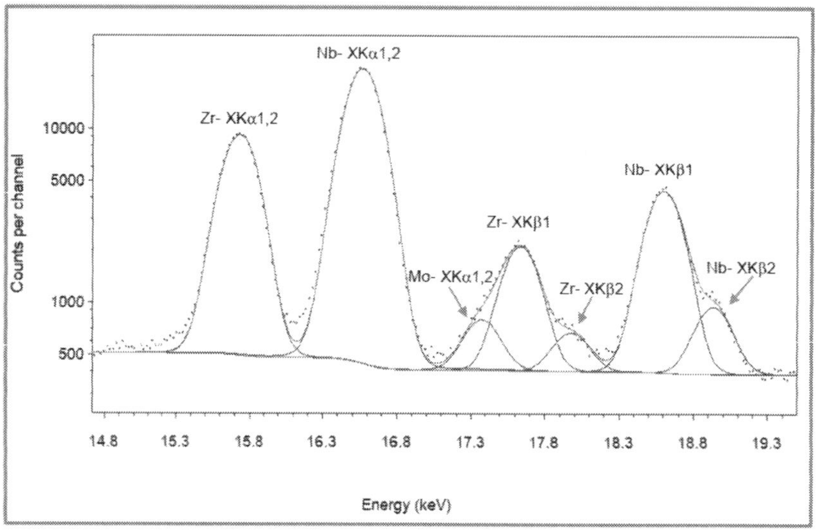

Fig. 3. Processing of the X-ray energy region of interest using COLEGRAM software

2.4 Fluorescence corrections

The main difficulty encountered in post-irradiation measurements is the presence of both niobium radioisotopes and the impurities in the dosimeter material, such as tantalum, which induce niobium fluorescence and contribute to increase counts in the niobium X-ray peaks.

2.4.1 Impurities main characteristics

Niobium-92m decays by electron capture and beta plus emission to the excited levels of zirconium-92; consequently, it also emits K X-rays of zirconium which are well identified in the spectra (see e.g. Fig. 3).
Niobium-94 and niobium-95 decay by beta minus emission to the excited levels of molybdenum-94 and molybdenum-95, respectively. Both provide K X-rays of molybdenum that are well separated from the niobium ones in the X-ray spectra. The molybdenum XKβ emission (19.62 keV) being just above the niobium K ionization threshold (18.98 keV), it has a high probability of photoelectric interaction in the dosimeter material and thus of generating niobium fluorescence. However, the low intensity of Mo K X-ray emissions (about 0.2 % and 0.01 % for ^{94}Nb and ^{95}Nb respectively) limits their effect.
Niobium-95m decays mainly (97.5 %) by gamma transition to the niobium-95 ground level and 2.5 % by beta minus emission to the excited levels of molybdenum-95. It is characterized by its gamma emission at 235.69 keV (25.1 %) and K X-rays of niobium following electronic rearrangement after ejection of conversion electrons. Tantalum-182 decays by beta minus emission to the thirteen excited levels of tungsten-182. These rearrange in about 40 gamma transitions, with energies between 32 keV and 1453 keV. At low energy, one notes an intense gamma emission at 67.75 keV (43.6 %) and the K X-rays of tungsten at 58.8 keV and 67.7 keV, with respective intensities of 27 % and 7 %.
The decay scheme of ^{183}Ta is similar to that of ^{182}Ta, decaying by beta minus emission to eleven excited levels of tungsten-183, which give about 30 gamma transitions. The gamma energies are below 406 keV, the main ones being 246 keV and 354 keV with intensities of 27 % and 11.2 % respectively; the intensities of the K X-ray emissions from tungsten being of the order of 68 % for XKα and 18 % for XKβ.

2.4.2 Fluorescence effect on X-ray spectra

Any photon radiation from the decay of radioactive impurities with an energy above the K-binding energy of niobium can interact by photoelectric effect in the atomic K shell of the dosimeter material. The atomic relaxation following photoionisation leads to the emission of K X-rays of niobium. These induced fluorescence photons are detected in the same way as the K X-rays characteristic of ^{93m}Nb decay, with the same energies and therefore contribute to increasing the areas of the full-energy peaks recorded in the spectra. It is therefore necessary to calculate factors to correct for this effect. This problem was studied in the doctoral work of Jonathan Riffaud [8] where the fluorescence induced by each impurity was simulated by Monte Carlo method in order to determine the different corrective factors.

3 Monte Carlo simulations

The fluorescence contribution from radionuclide N is obtained by simulating its radioactive decay in a niobium dosimeter, thus inducing the fluorescence of this material. This requires simulating the complete decay scheme of each radionuclide, taking into account interactions of the emitted particles with matter and atomic relaxation processes, using the PENNUC decay data file [9] from the Decay Data Evaluation Project (DDEP). The radioactive impurity was isotropically distributed inside the dosimeter and the simulation was performed under the geometrical measurement conditions of the X-ray spectrometer. The information collected corresponds to the energy deposited in the detector with the information of the process responsible for the niobium ionization. The simulated spectra include the characteristic peaks of the simulated radionuclide and the induced fluorescence peaks. Most of the correction factors for fluorescence were established by Monte Carlo simulation using GEANT4 [10]. In addition, the correction factor for ^{183}Ta was calculated with the PENELOPE 2018 code [11]. An example of the simulation result for ^{182}Ta is shown in Fig.4, where only the energy deposition is displayed (the experimental Gaussian broadening is not taken into account). In this low-energy spectrum, one can observe the peaks characteristic of the ^{182}Ta decay (gamma emissions and X-ray emissions from tungsten) as well as the peaks (in red) corresponding to the induced niobium fluorescence.

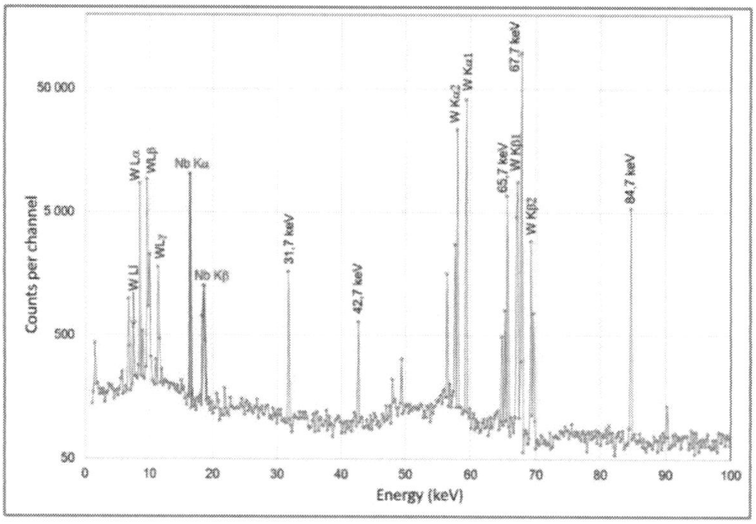

Fig. 4. Simulated spectrum following the decay of ^{182}Ta in a niobium dosimeter

The correction factor for the niobium K X-ray peak with energy *E*, is the number of events corresponding to this energy (the channel content minus the background), N_P, divided by the number of simulated radioactive decays, N_T, and by the X-ray spectrometer efficiency $\varepsilon(E)$ (Equation 2):

$$C_N(\mathrm{E}) = \frac{1}{\varepsilon(E)} \cdot \frac{N_P}{N_T} \qquad (2)$$

The correction factor was calculated for each of the niobium XKα and XKβ peaks and is expressed in $s^{-1}\cdot Bq^{-1}$; it corresponds to the number of X rays emitted per decay of the specific radioactive impurity. Therefore, it must be multiplied by the activity of the impurity and the detection efficiency to determine the spurious fluorescence count rate that must be subtracted from the net area of the full-energy peak. The correction factors for the different impurities are presented in Table 2; typical relative uncertainties are 1 % for the Kα corrective factors and 3 % for the Kβ ones.

Table 2. Correction factors for fluorescence induced by gamma-emitting impurities

Radionuclide	Corrective factor for Kα ($s^{-1}\cdot Bq^{-1}$)	Corrective factor for Kβ ($s^{-1}\cdot Bq^{-1}$)
^{92m}Nb	$1.489\cdot 10^{-4}$	$2.198\cdot 10^{-5}$
^{94}Nb	$1.003\cdot 10^{-2}$	$1.85\cdot 10^{-3}$
^{95}Nb	$2.55\cdot 10^{-3}$	$5.5\cdot 10^{-4}$
^{95m}Nb	0.364	0.0712
^{182}Ta	$5.42\cdot 10^{-2}$	$1.05\cdot 10^{-2}$
^{183}Ta	$8.21\cdot 10^{-2}$	$1.48\cdot 10^{-2}$

This study allows deriving some general remarks on the importance of the impurities:

i. The influence of ^{92m}Nb is very small, as it emits mainly energetic photons (934 keV); the K X-rays of zirconium are well separated on the spectra and are therefore not considered here.

ii. Both ^{94}Nb and ^{95}Nb emit K X-rays from molybdenum; since the energy of the molybdenum XKβ (19.62 keV) is just above the niobium K ionization threshold (18.98 keV), it has a high probability of photoelectric interaction in the dosimeter material and thus of generating the niobium fluorescence. However, the low intensity of Mo K X-rays emissions (about 0.2 % and 0.01 % for ^{94}Nb and ^{95}Nb respectively) limits their contribution.

iii. The effect of ^{95m}Nb is reflected both by the fluorescence induced by 200 keV photons and especially by the direct emission of K X-rays from niobium, leading to a large correction factor, but in many cases, ^{95m}Nb can be allowed to decay.

iv. The correction factors for ^{182}Ta and ^{183}Ta are of the same order of magnitude, the fluorescence being mainly induced by the K X-rays lines of tungsten.

In addition, with the information on the process responsible for the niobium ionization, it was shown that electrons are mainly responsible for the fluorescence effects in niobium dosimeters [8].

4 Experimental validation

To validate the calculated correction factors, these were applied to nine dosimeters (3 strips and 6 discs) with three impurity levels (0.3 ppm, 4 ppm and 19.6 ppm). These dosimeters were measured six times throughout the year following the end of the irradiation using gamma- and X-ray spectrometry to determine respectively the impurities activity and the Nb X-rays emission rates. These counting rates were corrected for the impurities contribution using the calculated correction before deriving the ^{93m}Nb activity at the same reference date. It must be noted that the very short-lived impurities (^{95m}Nb and ^{183}Ta) were not taken into account and that ^{92m}Nb only appears in the first series of measurements. Figure 5 shows the one-year evolution of the raw and corrected values for three among the six disc dosimeters.

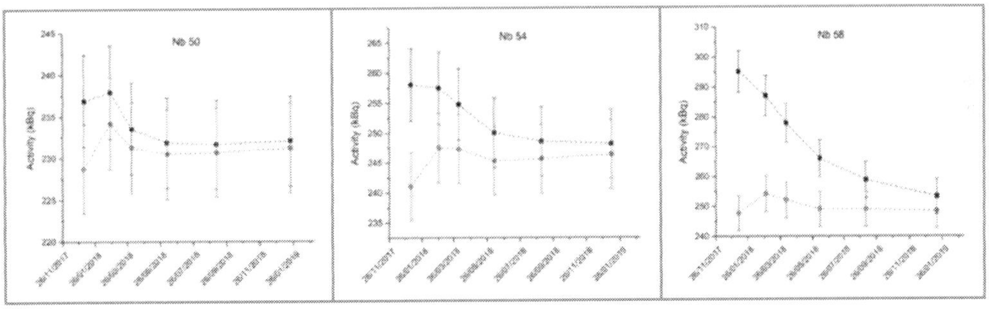

Fig. 5. Activity of ^{93m}Nb at the reference date for 3 discs without (black) and with (red) correction of the fluorescence of impurities (Ta impurities 0.3, 4 and 19.6 ppm)

In figure 5, the red curves of the activities obtained after application of the calculated correction coefficients are almost flat and thus show the relevance of the application of these coefficients, to obtain, whatever the cooling duration and the importance of the initial deviation, a reliable value of activity. These results are summarized in Table 3 that presents the mean values obtained during the six series of measurements, and compares the raw activity values (column 3) with the fluorescence-corrected values (column 6). The individual activities are obtained with a relative standard uncertainty of about 2.3 %. Maximum fluorescence correction factors (at the reference date) range from a few percent (3.5 %) for the 0.4 ppm Ta samples to about 17 % for the 19.6 ppm ones. It can be seen that taking into account the fluorescence corrections leads to consistent activity values at each measurement date: the standard deviation between the six measurements is reduced to about 1 %. At the same time, the mean value of the six results is significantly reduced, by up to 10 % for dosimeters containing a high proportion of tantalum.

Table 3. Mean activity values and associated standard deviation (SD) without and with fluorescence corrective factors, and relative differences between these.

Dosimeter number	Ta (ppm)	Mean raw activity (R) (kBq)	SD (kBq)	Relative SD (%)	**Mean corrected activity (C) (kBq)**	**SD (kBq)**	**Relative SD (%)**	**Relative difference (C-R)/R*100 (%)**
Nb8 (S)	0.3	23. 28	0.26	1.11	**22.86**	**0.45**	**1.98**	**-1.84**
Nb17 (S)	4	34.84	0.77	2.22	**33.88**	**0.42**	**1.23**	**-2.83**
Nb26 (S)	19.6	27.48	1.56	5.67	**25.39**	**0.40**	**1.90**	**-8.23**
Nb49 (D)	0.3	232.23	1.26	0.54	**229.35**	**2.68**	**1.17**	**-1.26**
Nb50 (D)	0.3	233.97	2.77	1.18	**231.08**	**1.77**	**0.77**	**-1.25**
Nb53 (D)	4	255.30	5.93	2.32	**247.88**	**0.74**	**0.30**	**-2.99**
Nb54 (D)	4	252.84	4.50	1.78	**245.52**	**2.34**	**0.95**	**-2.98**
Nb57 (D)	19.6	255.77	14.52	5.68	**234.33**	**1.31**	**0.56**	**-9.15**
Nb58 (D)	19.6	273.08	16.47	6.03	**250.13**	**2.55**	**1.02**	**-9.18**

5. Conclusion

The methodology based on Monte Carlo simulations to calculate fluorescence effects has been successfully applied to correct X-ray peaks areas from nine niobium dosimeters with different level of tantalum impurities. The fluorescence correction factors determined by calculation were applied to these irradiated dosimeters and allowed to obtain consistent results between the different measurements carried out over a one-year period. It can be concluded that, using this approach, it is possible to perform reliable measurements by X-ray spectrometry shortly after the activation date. However, it must be noted that the correction factors are geometry dependant (dosimeter thickness and composition, detection efficiency), therefore specific correction factors should be computed for each experimental condition.

References

1. ASTM E1297, Standard Test Method for Measuring Fast-Neutron Reaction Rates by Radioactivation of Niobium (2018)
2. D. Thorton *et al.*, The Second EWGRD Round Robin: Inter-Comparison of ^{93m}Nb Measurements (this conference)
3. IAEA. ENDF: Evaluated Nuclear Data File, http://www-nds.ciae.ac.cn/exfor/endf.htm (2018)
4. DDEP, http://www.lnhb.fr/ddep_wg/ (2022)
5. M.-C. Lépy, A. Pearce and O. Sima, Metrologia 52, S12-145 (2015), http://dx.doi.org/10.1088/0026-1394/52/3/S123
6. Y. Ménesguen, M.-C. Lépy, Nucl. Instrum. Methods Phys. Res. A, 1003, 165341 (2021), https://doi.org/10.1016/j.nima.2021.165341

7. F. Piton, M.-C. Lépy, M-M. Bé, J. Plagnard, Appl. Radiat. Isot., 52, 791-795 (2000), https://doi.org/10.1016/S0969-8043(99)00246-8
8. J. Riffaud, PhD thesis, Université Paris-Saclay (2018)
9. E. García-Toraño, V. Peyrés, M.-M. Bé, C. Dulieu, M.-C. Lépy, and F. Salvat., Nucl. Instrum. Methods Phys. Res. B, 396, 43-49 (2017)
10. S. Agostinelli, Nucl. Instrum. Methods Phys. Res. A, 506, 250–303 (2003)
11. F. Salvat, PENELOPE-2018: A code System for Monte Carlo Simulation of Electron and Photon Transport, Document NEA/MBDAV/R(2019)1, OECD Nuclear Energy Agency, Boulogne-Billancourt, France, 2019. Available from http://www.nea.fr/lists/penelope.html.

Design Considerations for Boron-Diffused and Implanted 4H-SiC Epitaxial Neutron Detectors for Dosimetry and Monitoring Applications

Frank H. Ruddy[1,*], *Joshua W.* Kleppinger[2], and *Krishna C.* Mandal[2]

[1]Ruddy Consulting, 2162 Country Manor Dr., Mount Pleasant, South Carolina, USA
[2]Department of Electrical Engineering, University of South Carolina, Columbia, South Carolina, USA

Abstract. Thermal-neutron detectors based on 4H-SiC semiconductor and ^{10}B converter reactions have many advantages for neutron dosimetry and monitoring applications. These radiation-resistant detectors are capable of stable operation in elevated-temperature environments up to 700 °C for extended periods. The recent development of SiC detectors where the ^{10}B is incorporated into the detector by ion implantation or diffusion leads to interesting application-specific design possibilities. The design of boron-diffused detectors is discussed as well as ways to optimize their design.

1 Introduction

Neutron detectors based on Silicon Carbide (SiC) have many advantages for neutron dosimetry and monitoring applications [1]. Following the first demonstration of radiation detectors based on 4H-SiC epitaxial layers [2], rapid progress and refinement of SiC detectors have been achieved [3-7]. SiC neutron detectors are particularly useful for measurements in the high-temperature, high-radiation environments typically encountered in nuclear power applications [8]. The wide bandgap (3.27 eV at 300K) results in far fewer thermally generated charge carriers compared to conventional lower bandgap semiconductors such as silicon or germanium and allows operation in elevated-temperature environments up to 700 °C without the need for cooling. Furthermore, SiC radiation detectors are extremely radiation tolerant, and have been shown to be operational after gamma-ray doses up to 22.7 MGy (^{137}Cs) [9] and fast-neutron (E >1MeV) fluences up to 1.7×10^{17} cm^{-2}[10]. SiC detectors are particularly well suited to spent fuel monitoring applications as reported at previous symposia in this series [11-13].

2 Background

Like all neutron detectors, SiC detectors rely on detection of neutron-induced reaction products in the form of energetic ions, which can produce detectable ionization when they interact with the detector. SiC detectors based on boron neutron-convertor layers that have been implanted into the detector [14] or diffused into the detector [11] have been reported

* Corresponding author: FrankHRuddy@gmail.com

previously in this symposium series. In both cases, neutron-induced $^{10}B(n,\alpha)^{7}Li$ reaction products enter the active volume of the detector and produce ionization and detectable charge pulses.

The calculated [15] ranges in SiC for the ions produced in thermal neutron-induced ^{10}B reactions are summarized in Table 1. Thermal neutron reactions with ^{10}B atoms located

Table 1 Calculated Ranges in Silicon Carbide for ^{10}B Neutron-Induced Reaction Products

Particle	Energy (keV)	Range in SiC*
$^{10}B(n,\alpha)^{7}Li$ (480-keV excited state – 94%)		
α	1470	3.4 μm
^{7}Li	840	1.6 μm
$^{10}B(n,\alpha)^{7}Li$ (ground state – 6%)		
α	1777	4.14 μm
^{7}Li	1015	1.81 μm

more than 4.14 μm from the active volume of the detector are not detectable. The thermal-neutron induced reaction products from ^{10}B atoms located closer than 1.6 μm to the active volume are capable of reaching the active volume. If a sufficient amount of ionization is produced within the active volume, a signal corresponding to the neutron-induced reaction will be detected.

In the case of boron-implanted neutron detectors, a narrow layer of boron atoms will be produced at a depth corresponding to the range of the implanted boron atoms in the detector. Issa, *et al.* [14,16-18] have reported results for neutron detectors where 180-keV and 2-MeV boron ions were implanted in 2-μm and 4-μm aluminum contact layers, respectively, of SiC p-n diodes. The calculated projected ranges of the 180-keV and 2-MeV boron ions are 0.45 μm and 2.5 μm, respectively. Because the location of the boron atoms can be precisely controlled relative to the location of the detector active volume, it is possible to optimize the corresponding detector response. Detailed discussions of the detector-design issues and neutron-response results can be found in references [14] and [16-18].

In the case of boron-diffused detectors [11], boron (a p-type dopant) is diffused into n-type 4H-SiC to produce a p-n junction as depicted in Fig. 1. Under the influence of a reverse bias, the n-region is depleted of charge carriers and becomes the active volume of the detector as shown in Fig. 2. For the detector tested in reference [11], the thickness of the initial n-layer was 50 μm. Therefore, the thickness of the active volume of the fully depleted detector is 50 μm minus the distance of the p-n interface from the surface of the detector. Thermal-neutron interactions with ^{10}B atoms in the detector can produce energetic ^{4}He and ^{7}Li ions which produce ionization in the detector active volume as shown in Fig. 3. In the case shown in Fig.3, the location of the neutron reaction results in all of the energy of the ^{4}He and ^{7}Li being deposited within the active volume of the detector. Less energy may be deposited if the reaction occurs within the active volume but less than the range of either the ^{4}He or ^{7}Li ion from a boundary of the active volume or if the reaction occurs outside of the active volume. No ionization is expected to be produced in the detector active volume by $^{10}B(n,\alpha)^{7}Li$ reactions that originate from more than about 4.14 μm outside of the active volume. Therefore, if the peak boron concentration shown schematically in Fig. 3 is more than 4 μm from the p-n interface, the maximum ^{10}B concentration will not contribute to the detector response signal. Ideally, the peak boron concentration should be as close as possible to the p-n interface and not have a concentration that greatly exceeds the n-dopant (nitrogen)

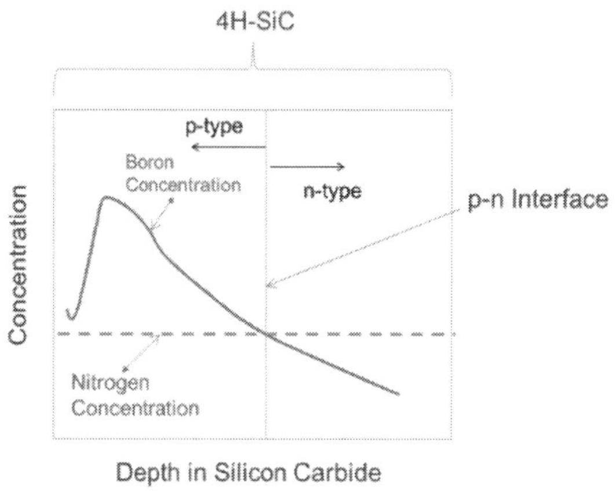

Fig. 1. Schematic representation of the regions of a p-n diode formed by diffusing boron into n-type SiC.

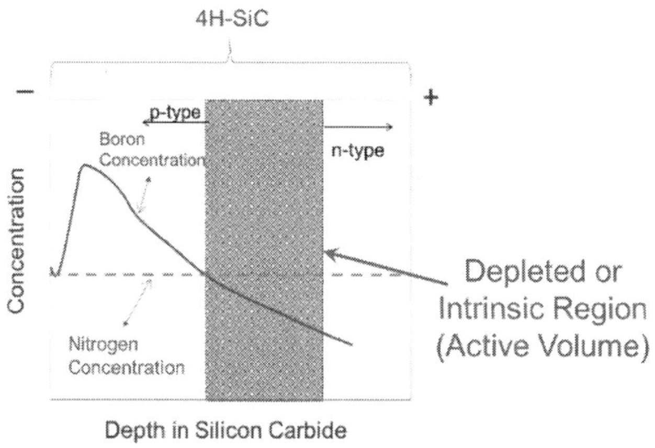

Fig. 2. Application of a reverse bias produces a depletion region, which acts as the active volume for the detector. A partially depleted detector is shown. The maximum depletion depth corresponds to the full thickness of the initial n⁻ region (distance from the p-n interface to the remainder of the epitaxial layer).

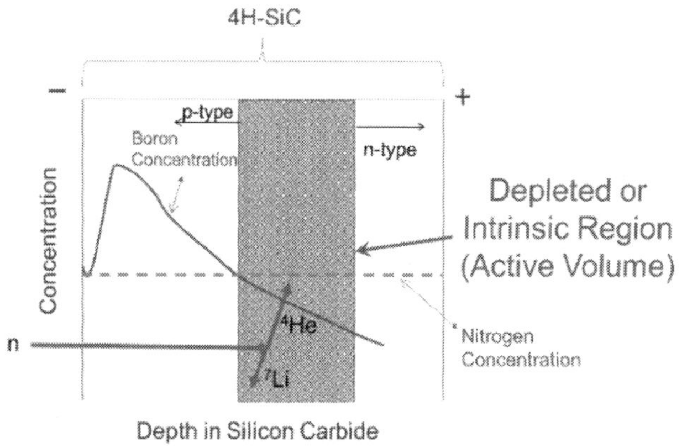

Fig. 3. A thermal neutron can interact with a ^{10}B nucleus within or adjacent to the depleted region to form energetic ^{4}He and ^{7}Li ions, which will produce ionization in the detector active volume.

concentration to avoid excessive neutron attenuation by ^{10}B that is not contributing to the detector response signal.

Therefore, the location and characteristics of the active volume and the detector thermal-neutron sensitivity depend upon several factors:

1. The amount of ^{10}B present.
2. The distribution profile of the boron in the SiC p-n diode.
3. The distance of each ^{10}B atom from the detector active volume.
4. The location and dimensions of the active volume.
5. The extent of attenuation of the thermal neutrons due to the presence of the ^{10}B.

An analysis of the influence of these factors on the observed response of a boron-diffused p-n diode thermal neutron detector follows.

3 The Thermal-Neutron Response of Boron-Diffused SiC Detectors

Fabrication and neutron-response testing of a boron-diffused SiC p-n diode detector were reported at the previous symposium in this series [11]. Secondary Ion Mass Spectrometry (SIMS) measurements of the boron concentration as a function of depth in SiC were reported for the 50-μm detector tested in reference [11]. Two components to the boron concentration profile were observed. A sharp peak with a boron concentration of about (3.5-4.7) x 10^{19} cm^{-3} at a depth of 0.2 μm from the surface was followed by a rapid drop of about a factor of fifty at a depth of approximately 0.26 μm after which a more gradual exponential decrease in concentration occurred up to a depth of 0.8 μm where the measured concentration was (3.9-5.8) x 10^{17} cm^{-3}. Measurement results were not obtained beyond 0.8 μm from the SiC surface. A fit to the boron-concentration data from reference [11] is shown in Figs. 4(a) and 4(b). It can be seen that the boron concentration at a depth of 0.8 μm exceeds the net n-carrier (nitrogen) concentration (~2 x 10^{14} cm^{-3}) by more than a factor of 2000. If a logarithmic extrapolation of the boron concentration is assumed as shown in Fig. 4 (a), the location where the boron and nitrogen concentrations are equal can be estimated as shown in Fig. 4 (b). It can be seen that the boron concentration exceeds the nitrogen concentration up to a depth of 4.8 μm, which corresponds to the location of the p-n interface in the SiC. From the data in Table 1, the maximum $^{10}B(n,\alpha)^7Li$ reaction product range is 4.14 μm, and as indicated in Fig. 4 (b), most of the boron atoms in the boron-concentration profile (including the peak) are located where interaction of the neutron-induced reaction products with the detector active volume corresponding to the detector n-region is not possible.

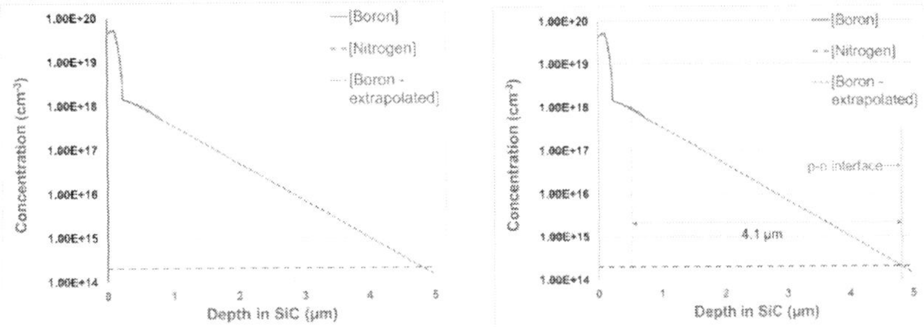

Fig. 4. (a) Measured [11] and extrapolated diffused-boron concentrations in SiC compared to the nitrogen-dopant concentration. **(b)** The p-n interface corresponds to the location where the boron and nitrogen concentrations are equal.

In addition to not effectively contributing to the SiC detector neutron response, the presence of the large concentration of ^{10}B atoms beyond the reaction-product range from the detector active volume will decrease the detector thermal-neutron response due to the high thermal-neutron cross section of ^{10}B (3838 barns) and resulting neutron attenuation.

Also, the extraneous $^{10}B(n,\alpha)^{7}Li$ reactions products will produce radiation damage over time which will reduce the detector service lifetime in high neutron fluence-rate applications.

Therefore, although the boron-diffused SiC detector reported in reference [11] produced a robust response signal, its design was far from optimal. Although it seems counter-intuitive, the first step in optimizing the detector response is to decrease the concentration of the diffused boron.

4 Optimization of the Thermal-Neutron Response of Boron-Diffused SiC Detectors

Initially, an overall reduction in the total boron present by a factor of 120 was assumed. The shape of the boron concentration distribution has been assumed to be unchanged as shown in Fig. 5. The depth of the resulting p-n interface is now 2.4 μm.

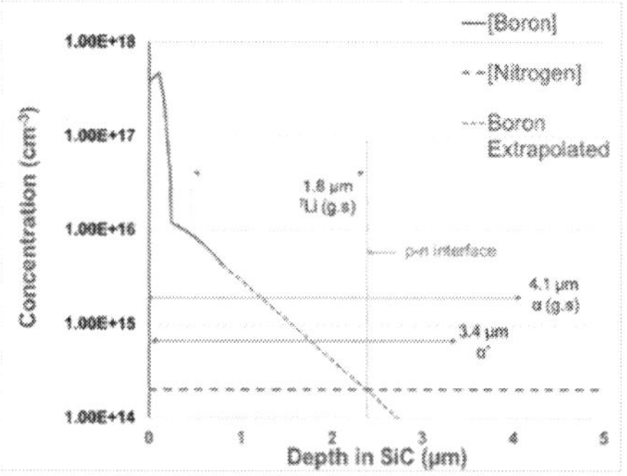

Fig. 5. Boron concentration distribution in SiC assuming a reduction of all boron concentrations by a factor of 120 from those used in the as-built detector configuration.

It can be seen from Fig. 5 that the entire boron concentration is located close enough to the p-n interface to allow the alpha particles from both the ground state (4.1-μm range) and excited state (3.4-μm range) of the $^{10}B(n,\alpha)^{7}Li$ reaction to reach the detector active volume. However, most of the 1.8-μm ^{7}Li ions produced in the $^{10}B(n,\alpha)^{7}Li$ ground-state branch and the 1.6-μm ^{7}Li ions produced in the excited-state branch are not capable of crossing the p-n interface into the detector active volume, because most of the boron-concentration is located within 0.6 μm of the detector surface. Nevertheless, the sensitivity of this reduced boron-concentration detector configuration can be expected to be higher than that of the detector configuration reported in reference [11].

In order to assess the potential neutron sensitivities of the different detector configurations, a crude, two-dimensional model was developed where the sensitivity of the detector was taken to be proportional to a line integral of the products of boron concentration at points along the boron concentration distribution times the corresponding solid angles defined by the range of particles capable of reaching (or passing) the detector p-n interface. Isotropic reaction-product distributions were assumed. It was also assumed that a residual range of at least 1 μm would be needed to produce a sufficient ionization pulse in the detector active volume, and initially, only the 94% excited-state reaction branch was modeled.

The calculations indicate that approximately 0.0235% of the $^{10}B(n,\alpha)^{7}Li$ reactions can contribute to the signal in the detector as fabricated [11], but that roughly 28.6% can contribute if the concentration is reduced by a factor of 120. Therefore, a net <u>increase</u> in detector neutron sensitivity of roughly a factor of 10 should result if the boron content of the detector is reduced by a factor of 120. This increase results from inclusion in the detector response of more reactions from near the peak of the boron concentration distribution, whereas the original detector response relied on reactions from the low-concentration tail of the distribution.

In order to further increase the detector sensitivity by including the response to ^{7}Li ions and more of the α particles, a configuration where the original boron concentration was reduced by a factor of 1000 was assumed as shown in Fig. 6. It can be seen that the factor

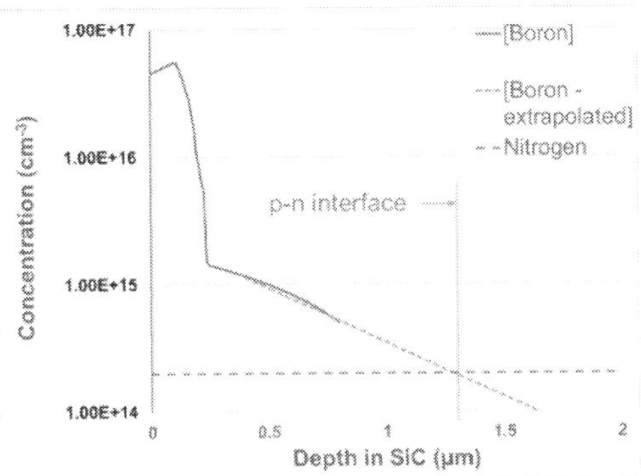

Fig. 6 Boron concentration distribution in SiC assuming a reduction of all boron concentrations by a factor of 1000 from those used in the as-built detector configuration.

of 1000 decrease in the boron concentration distribution has shifted the location of the p-n interface to 1.29 μm. Most of the ^{7}Li ions produced by neutron reactions with the boron distribution are now capable of crossing the p-n interface as are most of the α particles. The calculated fraction of the ^{10}B atoms that can contribute to the signal for this configuration is 68.6%. Therefore, reduction of the boron concentration distribution by a factor of 1000 results in a factor of 2900 times more detector sensitivity or a net sensitivity gain of a factor of 2.9.

The p-n detector configuration resulting from reducing the original boron concentration profile by a factor of 2800 results in a p-n interface that has been shifted further downward to 0.75 μm and the corresponding fraction of the boron concentration that contributes to the detector response increases to 90.1%. The decrease of a factor of 2800 in boron concentration has resulted in a factor of 3800 increase in detector sensitivity, or a net gain of a factor of 1.4.

The p-n detector configuration resulting from reducing the original boron concentration profile by a factor of 5600 results in a p-n interface that has been shifted further downward to 0.41 μm and the corresponding fraction of the boron concentration that contributes to the detector response has increased to 94.8%. The decrease of a factor of 5600 in boron concentration has resulted in a factor of 4830 increase detector sensitivity to the boron present, or a net gain of a factor of 0.7, *i.e.,* there is a net decrease in detector sensitivity by about 30%. Presumably, further decreases in the boron concentration would result in further decreases in net sensitivity, because the detector sensitivity is reaching its theoretical limit of 100% (2π geometry times two ionizing particles per neutron reaction). Therefore, further boron reductions will result in a proportional reduction in neutron sensitivity.

5 Conclusions

The results are summarized in Table 2. It must be emphasized that these results are based on a crude, two-dimensional model, and more precise results should use a more sophisticated three-dimensional model that accounts more precisely for the details of the boron-concentration distribution. In order to construct such a model, more detailed data are needed for the boron concentrations measured to date as well as information on the net n⁻ carrier (nitrogen) depth profile. Also, measurements that probe the location of the p-n interface and the depletion depth as a function of reverse bias are needed. In addition, the more detailed model should consider the thermal-neutron attenuation effects of the boron content of the detector on neutron sensitivity.

Table 2. Relative SiC p-n Detector Thermal-Neutron Sensitivities as a Function of Amount of boron Diffused into the Detector

Boron Concentration Distribution	p-n Interface Depth	Percent of Total Boron Contributing to Response	Relative Boron Contribution	Net Relative Increase in Response
Base Case (x1)	4.8 μm	0.0235%	1.00	---
Concentration/120	2.4 μm	28.6%	1220	10.1
Concentration/1000	1.29 μm	68.6%	2920	2.9
Concentration/2800	0.75 μm	90.1%	3830	1.4
Concentration/5600	0.41 μm	94.8%	4030	0.7

Nevertheless, the conclusions obtained with the crude detector-response model demonstrate the trade-off between amount of boron diffused into the detector and detector neutron sensitivity. Clearly, boron concentrations significantly less than originally used [11] are sufficient to produce optimal detector response. Based on the results reported herein, a more optimum detector response will likely be obtained with about a factor of 2000±500 reduction in the amount of boron used.

Boron-diffused 4H-SiC detectors are a promising technology for optimized, application-specific designs of sensitive, radiation-resistant detectors for monitoring and dosimetry in high-temperature environments.

References

1. F. Franceschini and F. H. Ruddy, "Silicon Carbide Neutron Detectors," Chapter 13 in *Properties and Applications of Silicon Carbide*, R. Gerhartd (Ed.), Intech, 2011, pp. 275-296 2011.
2. F. H. Ruddy, A. R. Dulloo, J. G. Seidel, S. Seshadri, and L. B. Rowland, "Development of a Silicon Carbide Radiation Detector," IEEE Trans. Nucl. Sci., **45**, pp. 536-541 1998.
3. F. Nava, G. Bertuccio, A. Cavallini, and E. Vittone, "Silicon Carbide and Its Use as a Radiation Detector," Meas. Sci. Technol., **19**, pp. 102001-1–25 2008.
4. F. H. Ruddy, "Silicon Carbide Radiation Detectors: Progress, Limitations and Future Directions," Materials Research Society Symposium Proceedings, **5761**, pp. 1-12 2013.
5. K. C. Mandal, J. W. Kleppinger, and S. K. Chaudhuri, "Advances in High-Resolution Radiation Detection Using 4H-SiC Epitaxial Layer Devices," Micromachines, 11, 254-1-27 2020.
6. S. K. Chaudhuri and K. C. Mandal, "Radiation Detection Using n-Type 4H-SiC Epitaxial Layer Surface Barrier Detectors," In: Iniewski K. (eds) Advanced Materials for Radiation Detection. Springer, Cham., Chapter 9, pp. 165-182, 2022. https://doi.org/10.1007/978-3-030-76461-6_9 .
7. J. W. Kleppinger, S. K. Chaudhuri, O. Karadavut, and K. C. Mandal, "Role of deep levels and barrier height lowering in current-flow mechanism in 150 μm thick epitaxial n-type 4H-SiC Schottky barrier radiation detectors," Appl. Phys. Lett., **119**, 063502-1-6 2021.

8. F. H. Ruddy, L. Ottaviani, A. Lyoussi, C. Destouches, O. Palais and C. Reynard-Carette, "Silicon Carbide Neutron Detectors for Harsh Nuclear Environments: A Review of the State of the Art," *IEEE Trans. Nucl. Sci.*, **69**, pp. 792-803 2022.
9. F. H. Ruddy and J. G. Seidel, "Effects of Gamma Irradiation on Silicon Carbide Semiconductor Radiation Detectors," IEEE Nuclear Symposium Conf. Record Paper 4179062, pp583-587, 2006.
10. F. H. Ruddy, A. R. Dulloo, and J. G. Seidel, "Study of the Radiation Resistance of Silicon Carbide Radiation Detectors," Trans. Am. Nucl. Soc. **90**, 348, 2004.
11. K. C. Mandal, T. A. Chowdhury, C. Oner, and F. H. Ruddy, "Design and Response Testing of Boron-Diffused Silicon Carbide Neutron Detectors for Dosimetry and Monitoring Applications," in *Reactor Dosimetry: 16th International Symposium*, ASTM STP1608, M. H. Sparks, K. R. Depriest, and D. W. Vehar, Eds., ASTM International, West Conshohocken, PA, 2018, pp. 353–360, http://dx.doi.org/10.1520/STP160820170042
12. A. R. Dulloo, F. H. Ruddy, J. G. Seidel, T. Flinchbaugh, C. Davison, and T. Daubenspeck, "Neutron and Gamma-Ray Dosimetry in Spent-Fuel Radiation Environments, Using Silicon Carbide Semiconductor Radiation Detectors," in *Reactor Dosimetry: Radiation Metrology and Assessment,* ASTM STP 1398, J. G. Williams, D. W. Vehar, F. H. Ruddy, and D. M. Gilliam Eds., 2001, pp. 683-90.
13. T. Natsume, H. Doi, F. H. Ruddy, J. G. Seidel, and A. R. Dulloo, "Spent Fuel Monitoring with Silicon Carbide Semiconductor Neutron/Gamma Detectors, " *Journal of ASTM International*, Vol. 3, Online Issue 3, 2006.
14. F. Issa, V. Vervisch, L. Ottaviana, D. Szalkai, L. Vermeeren, A. Lyoussi, A. Kuznetsov, M. Lazar, A. Klix, O. Palais, and A. Hallén, "Improvements in Realizing 4H-SiC Thermal Neutron Detectors," *ISRD 15 – International Symposium on Reactor Dosimetry,* A. Lyoussi (Ed.*),* EPJ Web of Conferences, **106**, 05004, 2016.
15. J. F. Ziegler, and J. P. Biersack. "SRIM-2013 software package," *see* http://www.srim.org (2013).
16. F. Issa, V. Vervisch, L. Ottaviani, D. Szalkai, L. Vermeeren, A. Lyoussi, A. Kuznetsov, M. Lazar, A. Klix, O. Palais, R. Ferone and A. Hallen, "Study of the Stability of 4H-SiC Detectors by Thermal Neutron Irradiation," Mater. Sci. Forum 821-823 pp. 875-878 2015.
17. F. Issa, V. Vervisch, L. Ottaviani, D. Szalkai, L. Vermeeren, A. Lyoussi, A. Kuznetsov, M. Lazar, A. Klix, O. Palais and A. Hallen, "Radiation Silicon Carbide Detectors Based on Ion Implantation of Boron," IEEE Trans. Nucl. Sci. **61**, 2105-2111 2014.
18. F. Issa, L. Ottaviani, D. Szalkai, L. Vermeeren, V. Vervisch, A. Lyoussi, R. Ferone, A. Kuznetsov, M. Lazar, A. Klix and O. Palais, "4H-SiC Neutron Sensors Based on Ion Implanted ^{10}B Neutron Converter Layer," IEEE Trans. Nucl. Sci. **63**, 1976-1980 2016.

Fabrication and characterization of high-resolution 4H-SiC epitaxial radiation detectors for challenging reactor dosimetry environments

Krishna C. Mandal[1], *Sandeep K.* Chaudhuri[1], *and Frank H.* Ruddy[2,*]

[1]Department of Electrical Engineering, University of South Carolina, Columbia, SC 29208, USA
[2]Ruddy Consulting, 2162 Country Manor Dr., Mount Pleasant, SC 28466, USA

Abstract. Reactor dosimetry environments require radiation detectors that are capable of operating at high temperatures in extremely high neutron and gamma-ray dose rates. Silicon carbide (SiC) is one of the most promising wide bandgap semiconductors (3.27 eV) for harsh environment applications due to its radiation hardness, high breakdown voltage, high electron saturation velocity, and high thermal conductivity. In this paper, we summarize the prospect of Schottky barrier radiation detectors, fabricated on highly crystalline low-defect detector-grade n-type 4H-SiC epitaxial layers with thickness ranging from 20 to 250 μm, for harsh environment applications. A comprehensive discussion on the characterization of the parameters that influence the energy resolution has been included. The usage of electrical and radiation spectroscopic measurements for characterizing the junction and rectification properties, minority carrier diffusion lengths, and energy resolution has been elaborated. Characterization of crucial factors that limit the energy resolution of the detectors such as charge trap centers using thermally stimulated transient techniques is summarized. Finally, the effect of neutron fluence on the performance of the 4H-SiC detectors is discussed.

1 Introduction

The advantages of nuclear power generation outweigh any renewable or non-renewable sources by a substantial margin provided the safety of nuclear reactors are assessed regularly, if not continuously [1] [2]. For example, reactor pressure vessels (RPV) in nuclear reactors are subjected to continuous bombardment by radiation from fission reactions of the nuclear fuel. Fast neutrons and, to a lesser extent, high energy gamma rays are primary causes of radiation damage as they cause embrittlement of the reactor vessel material with prolonged exposure and, hence, can lead to deterioration of the structural integrity of RPVs [3]. Standard safety measures comprise monitoring the damage in representative steel (RPV material) specimens and the corresponding neutron/gamma dose using dosimetry. The detectors used for neutron dosimetry must be capable of withstanding the harsh reactor environment i.e., high radiation doses and elevated temperatures. When it comes to radiation detection and

* Corresponding author: frankhruddy@gmail.com

measurements, semiconductor detectors have many advantages because of their compactness, fast and direct readout, high efficiency, and high resolution. However, most of the matured semiconductor detector technologies like silicon or germanium are not suitable for prolonged use in harsh environments due to their lower bandgaps and displacement thresholds. Lower bandgap causes high leakage currents even at room temperature and lower displacement threshold implies formation of a higher number of radiation-induced defects at lower energies. The 4H polytype of silicon carbide is a wide bandgap (3.27 eV at 300K) semiconductor that is extremely radiation hard due to the high displacement threshold of Si and C in the 4H-SiC lattice [4]. The thermal conductivity of 4H-SiC is also very high and its physical properties barely change with increased temperature. A 4H-SiC based JFET has been demonstrated to work continuously for prolonged periods at an operating temperature of 800 – 1000 °C [5] [6]. The last decade has seen tremendous improvement in the growth technology to produce highly crystalline 4H-SiC epitaxial layers using chemical vapor deposition. These layers demonstrate excellent charge transport properties and yet cost much less to produce than single crystalline electronic-grade diamond [7]. Silicon carbide is also known for its physical strength, and it is also not prone to chemical corrosion. Being a wide bandgap semiconductor, 4H-SiC is insensitive to visible light and does not require light-tight encapsulation for most applications. Also, 4H-SiC is the only wide bandgap semiconductor whose native oxide is SiO_2, making it compatible with the matured silicon device fabrication technology. The unique combination of the above properties makes 4H-SiC the ideal and future material for electronic devices in harsh environments [8]. 4H-SiC epitaxial layers have been demonstrated as versatile high-resolution radiation detectors which are equally capable of detecting energetic charged particles as well as penetrating radiation such as neutrons and gamma rays [9] [10] [11] [12] [13] [14]. Fabrication of detectors for penetrating radiations has gained even more momentum recently with the advent of detector-grade thick (up to 250 μm) epitaxial layers [15] [16] [17]. In this article we discuss developments in the fabrication of state-of-the-art SiC Schottky barrier detectors (SBD) for harsh-environment applications and the techniques that predict their operation under such conditions through various material and device characterization techniques. The effect of fast-neutron bombardment on these radiation detectors is also discussed.

2 SiC epilayer growth and device fabrication

Growth of high-quality 4H-SiC epitaxial layers with a high degree of crystallinity is usually achieved through hot-wall chemical vapor deposition (CVD) [18]. The epitaxial layers discussed in this article are grown by hot wall CVD using dichlorosilane (SiH_2Cl_2) and propane (C_3H_8) as the precursors and hydrogen as the carrier gas. Unintentional doping of nitrogen impurity atoms results in n-type conductivity of the epilayers. The key factors to maintain during the growth are the carbon-to-silicon ratio which is necessary to control the formation of native point defects such as carbon/silicon vacancy/interstitials and other defect complexes, such as carbon antisite vacancy (CAV) pairs, as well as the growth rate which determines the uniformity of the crystalline nature of the epilayers. While impurity type defects such as titanium are mostly shallow levels, intrinsic deep level trap centers such as carbon vacancies pose serious threats as charges trapped in deep levels are prone to recombination leading to loss of signal and degradation of detector performance [4]. Apart from point defects, extended structural defects such as threading screw dislocations (TSD), threading edge dislocations (TED), and basal plane dislocations (BPD) also affect the crystallinity of the epilayers [19]. While TEDs are considered benign, micropipe defects (MPD), which are a type of TSDs, form high leakage current paths. The epilayers used in this study showed the presence of MPDs in areal concentrations less than 1 cm^{-2}. This low density enables the fabrication of detectors with entrance windows having large surface areas, because of the low probability of subtending a defect. To prevent the formation of extended defects and their propagation from the substrate to the epilayer during the growth, off-axis, step-controlled growth techniques are adopted, and thin buffer layers are included. The present epilayers, with thickness ranging between 20-250 μm, are grown on the (0001) plane (Si face) of highly conducting n-type 350-μm thick substrates, 8° offcut towards the $\langle 11\bar{2}0\rangle$ direction. A 1 μm thick buffer epilayer with a nitrogen doping concentration of 1×10^{18} cm^{-3} has been included between the bulk substrate and the epilayer. Fig. 1(a) shows a schematic diagram of the various crystal planes and Fig. 1(b) shows the epilayer structure. The epilayers grown on 100 mm diameter substrates are diced in 8 mm × 8 mm square specimens for detector fabrication. The wafers are cleaned thoroughly using a Radio Corporation of America (RCA) cleaning method which removes organic and metallic contaminants [20]. The wafers are treated with 50% diluted hydrofluoric acid for 30 secs to remove the native SiO_2 layer.

To form the Schottky barrier junction, which also acts as the detector entrance window, a nickel layer with thickness less than 20 nm is deposited on the epilayer through sputter coating. Such thin contacts form transparent windows for most radiations [21]. The detectors used in this study have circular nickel contacts with an area of 11 mm^2. A thicker (≈100 nm) nickel contact is deposited on the bulk side to form the Ohmic cathode. The detectors are generally mounted on printed circuit boards (PCB) using conductive epoxy on the cathode side and wire bonded on the anode for radiation detection measurements, However, for several other measurements spring loaded removable contact probes have been used. Fig. 1(c) shows a typical 4H-SiC SBD mounted on a PCB.

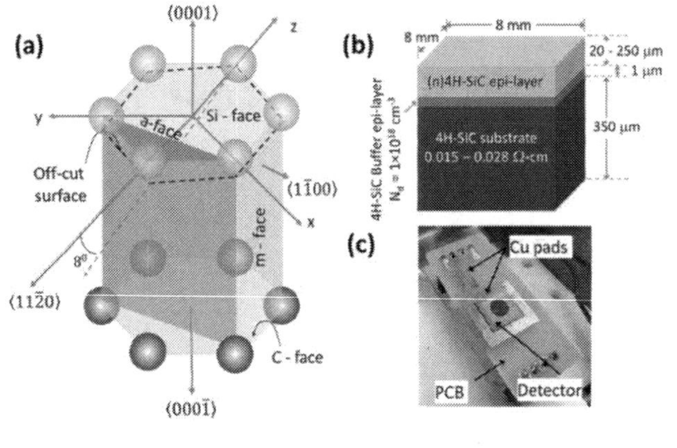

Fig. 1. Schematics showing the (a) crystal faces of 4H-SiC epilayers and (b) the substrate-epilayer configuration used for detector fabrication. (c) Photograph of a typical Ni/4H-SiC detector mounted on a PCB. The circular Ni Schottky and the square back contacts are also visible.

3 Device characterization

3.1 Electrical characterization

The performance of the SBDs depends on the material as well as the junction properties. The effective doping concentration of the epilayers used in this study is on the order of 10^{14} cm^{-3}, which is high enough to obscure the charge pairs (≈3×10^{9} cm^{-3}) generated by incoming radiations. Hence, to generate a detector signal due to the motion of radiation induced charge pairs only, a detector active region is needed where the equilibrium carrier concentration is orders of magnitude lower compared to the generated number of charge pairs. A Schottky barrier configuration is the most efficient and convenient way to generate a depletion region just underneath the contact surface, which is practically devoid of free carriers. Depending on the effective doping concentration of the 4H-SiC epilayer, depletion widths more than 100 μm can be achieved with a reverse bias of a few hundred volts. Puglisi *et al.* reported full depletion of a Schottky-type microstrip detector fabricated on a 124-μm thick 4H-SiC epilayer with a doping concentration of 5.2×10^{13} cm^{-3} at a bias voltage of 600 V [13]. The Schottky barrier height (SBH) restricts the leakage current flow under reverse bias which is important to reduce the electronic noise of the detector. Several refractory metals such as Ni, Ti, Pt, Re, W, Mo, and Ru form very stable Schottky contact which can withstand high temperatures normally encountered around RPVs and offer SBH over 1.5 eV ([22] and references therein). Apart from the current flow through the junction, charge carriers also flow on the surface which contributes to the total leakage current. Guard rings and/or passivated surfaces suppress the surface leakage current to a great extent. The current density J through the junction is mostly thermionic emission over a metal-semiconductor junction modelled as Eq. 1 [22].

$$J = A^{*}T^{2}\left(e^{-\beta\varphi_B}\right)\left(e^{\frac{\beta V}{n}} - 1\right), \tag{1}$$

where, A^* is the effective Richardson constant (146 Acm^{-2}K^{-2} for 4H-SiC), φ_B is the SBH, n is the diode ideality factor, V is the applied voltage, and $\beta = q/k_BT$, q being the electronic charge, k_B the Boltzmann constant, and T the absolute temperature. A $\ln J - V$ plot, V being the applied bias, is generated from the experimentally measured leakage current as a function of V. The SBH and the ideality factor are calculated from the intercept and slope of a linear fit of Eq. 1 to the $\ln J$- V plot[22]. Fig. 2(a) shows a $\ln J$- V plot in forward and reverse bias obtained for a Ni/20μm-4H-SiC SBD without a guard ring measured at room temperature using a Keithley 237 source-measure unit. The barrier height and the ideality factor were calculated to be 1.6 eV and 1.09, respectively. An ideality factor close to unity indicates that the barrier height is distributed uniformly across the contact area. Defects on the surface creates patches with low barrier heights through which a larger fraction of the total current may flow making the effective SBH lower [23]. The effective SBH and ideality factor also improves at high temperatures where the charge carriers acquire higher energy to overcome the higher SBH in the areas outside the low barrier patches [24]. The leakage current and ideality factor obtained for a similar but thicker Ni/150μm-4H-SiC SBD showed an improvement of the ideality factor from 1.9 at 350 K to 1.1 at 600 K with a corresponding increase in the SBH from 1.5 to 2.1 eV [16].

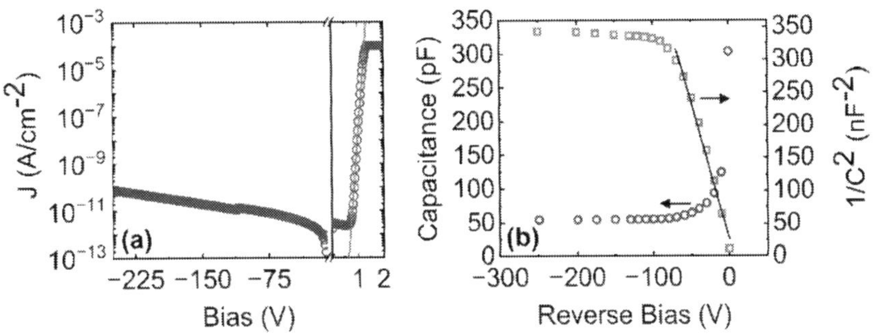

Fig. 2. (a) Semilogarithmic plot of current density as a function of forward and reverse bias for a Ni/20μm-4H-SiC SBD. The solid line is a linear fit of Eq. (1) to the forward *lnJ-V* plot. (b) Variation of detector capacitance (left y-axis) measured at 1 MHz test frequency as a function of reverse bias voltage and the corresponding Mott-Schottky plot (right y-axis).

Normally, the selection of the operating bias of radiation detectors is a trade-off between the leakage current and the charge collection efficiency (CCE). While it is essential to create an electric field high enough to ensure the collection of all the generated charge carriers, high fields can generate high leakage currents which increases the electronic noise. The Ni/4H-SiC detectors, however, allow 100% CCE and/or full depletion without substantial increase in the detector leakage current given the high barrier height and low effective doping concentration. The doping concentration is derived from the Mott-Schottky plots of the capacitance-voltage (C-V) characteristics at high measurement frequency [15]. Fig. 2(b) shows a C-V plot and the corresponding Mott-Schottky plot obtained for a Ni/20μm-4H-SiC SBD measured at a test frequency of 1 MHz. The operating bias voltage is usually set as the minimum bias at which the maximum CCE is observed. Further fine tuning of the bias is done to achieve the best resolution and will be discussed in Section 3.2.

3.2 Radiation detection

Alpha particle radiation detection measurements were carried out with a standard benchtop radiation spectrometer comprised of an Amptek A250CF preamplifier, an Ortec 576 shaping amplifier, a Canberra Multiport II MCA, and Genie 2000 - a PC based data acquisition program [15]. The detectors are exposed to a 1 μCi ^{241}Am radioisotope emitting primarily alpha particles with three energies *viz.*, 5486, 5442, and 5388 keV. The detector and the source were kept inside a noise-shielded test box, which was continuously evacuated during the measurements. The detector was biased using a low noise Canberra 3106D power supply. Fig. 3(a) shows the variation of energy resolution and CCE of a Ni/20μm-4H-SiC SBD with reverse bias, calculated for the 5486 keV peak. The energy resolution is defined as the full width at half maximum (FWHM) in keV units or as the percentage of the peak energy. The CCE is defined as the ratio of the energy detected by the detector to the energy incident on the detector. It is usually calculated by the ratio of the alpha peak position in the calibrated spectrum to the incident alpha energy which in this case is 5486 keV. The observed initial improvement in the detector resolution with bias is due to the decrease in detector capacitance [9]. The CCE also increased initially as expected, and saturated to 100% at around -70 V. The best resolution was however obtained for -90 V with a shaping time of 3 μsec. The shaping time is optimized to obtain the highest peak resolution for the optimized bias. The recorded pulse height spectrum under the optimised setting is shown in Fig. 3(b). All the three primary alpha particles emitted by the source were fully resolved. The FWHM of the 5486 keV peak was measured to be 16.2 keV or 0.29% [9]. The energy resolution is comparable to the silicon SBD available commercially.

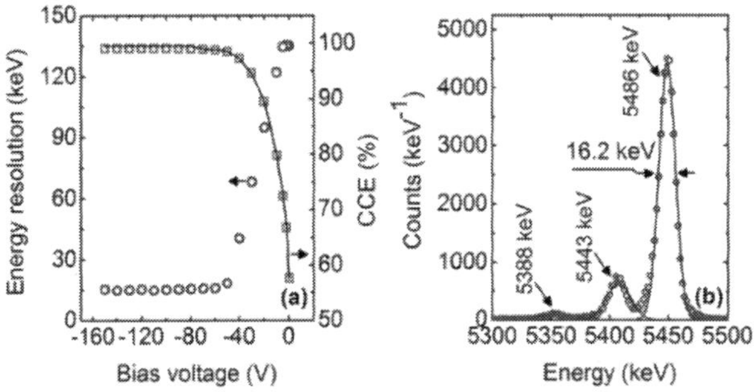

Fig. 3. (a) Variation of energy resolution (blue circles, left axis) and CCE (green squares, right axis) with reverse bias. The solid line shows the fitting of Eq. 2 to the measured CCE values. (b) Pulse height spectrum obtained using a Ni/20μm-4H-SiC SBD exposed to a ^{241}Am alpha source. The solid lines are the Gaussian fits.

3.3 Minority carrier diffusion length

The charge transport in a fully depleted detector, where all the radiation generated charge pairs are produced within the depletion region, is governed by the majority carriers. However, there are situations when the interactions occur beyond the depletion region (neutral region). In such cases the minority carriers (holes in the present n-type detectors) diffuse to the edge of the depletion region followed by a drift across the depletion width. The CCE in such a case is defined by the minority carrier transport (diffusion). A longer minority carrier diffusion length in such cases helps to achieve a 100% CCE even though the electrons are not collected. Such a situation is common for partially depleted detectors exposed to penetrating radiations such as during self-biased operations. Self-biased detectors are highly suitable for dosimetric applications where the logistics of locating a bias supply in or close to a high-temperature or high-radiation environment may by problematic [25]. Eq. 2 is a drift-diffusion model which expresses the variation of CCE (η) as a function of the detector bias (depletion width d) [26]. The solid line in Fig. 3(a) is a least square fitting of Eq. 2 to the variation of CCE with the bias voltage. The minority carrier diffusion length L_d was used as a fitting parameter and was found to be ~18 μm which is high enough to explain the 56% CCE at 0 V applied bias.

$$\eta = \frac{1}{E_p}\int_0^d \left(\frac{dE}{dx}\right) dx + \frac{1}{E_p}\int_d^{x_r} \left[\left(\frac{dE}{dx}\right) \times exp\left\{-\frac{(x-d)}{L_d}\right\}\right] dx \quad (2)$$

where, dE/dx is the electronic stopping power of the alpha particles calculated using SRIM 2018 for several energies up to the one needed by the alpha particles to penetrate a depth d inside the detector, and x_r is the range of the alpha particles with energy E_p[27].

4 Neutron irradiation effects

In the above discussions, the performance of the best 4H-SiC detectors has been dealt with. Although, ultra-low leakage currents and 100% CCE are routinely achieved using these detectors, the energy resolution of the detectors is still far from the theoretical limits. This is primarily due to the presence of electrically active point defects in the as-grown epilayers which act as charge trapping centers. A detailed study on the correlation between the role of point defects and detector performance can be found in Ref. [28]. It is only obvious that with the accumulation of radiation dose in a reactor environment, over a prolonged period of use of the detectors, new types of trapping centers evolve as well as the concentration of the intrinsic trap centers increases. The effects of fast neutron irradiation on 250 μm thick 4H-SiC epilayers at four different neutron fluences have been studied [29]. Ni/4H-SiC SBDs were fabricated following the irradiation. The epilayers were irradiated in a TRIGA Mk. II nuclear reactor dry instrumentation tube at Kansas State University at net neutron fluences of $5 \times (10^{10}, 10^{11}, 10^{12}, \text{and } 10^{13})$ cm^{-2} in addition to background gamma doses varying from 400 Rad – 300 kRad linearly with the neutron fluence. Fig. 4(a)-(d) shows the changes in the detector performance with increases in neutron fluence. The energy peaks corresponding to the 5486 keV alpha particles have shifted left with increased neutron fluence. With the increase in the neutron fluence it is likely that a detector has accumulated higher concentration of newer as well as the existing defects. These defects, increased in concentration, participate in increased trapping of the radiation induced charge pairs leading to increased charge loss. The peak position in the pulse height spectrum, being directly proportional to the number of charges collected by the detector, shifts to a lower channel number with increased concentration of charge trapping centers. The deterioration of the energy resolution with increasing neutron fluence is clearly visible as the lower energy (5443

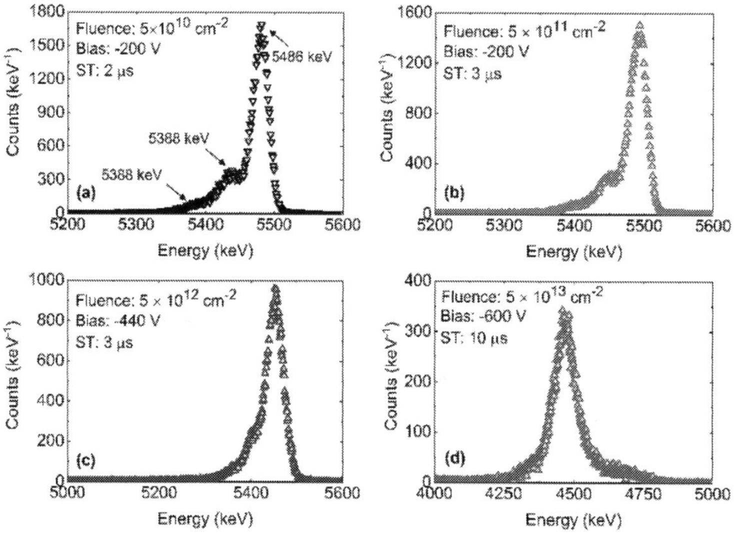

Fig. 4. (a)-(d) Pulse height spectra obtained for a Ni/250μm-4H-SiC SBD exposed to a ^{241}Am alpha source and irradiated with different neutron fluences.

and 5388 keV) alpha peaks gradually became unresolvable. Although the detectors with the first two doses showed similar energy resolution of ≈0.5% for the 5486 keV peak at -200 V, the one irradiated with higher fluence needed a larger shaping time of 3 μsec compared to the 2 μsec corresponding to the detector irradiated at the lowest dose. The increase in the optimum shaping time indicates increased trapping in the defects. Fig. 5(a) shows the

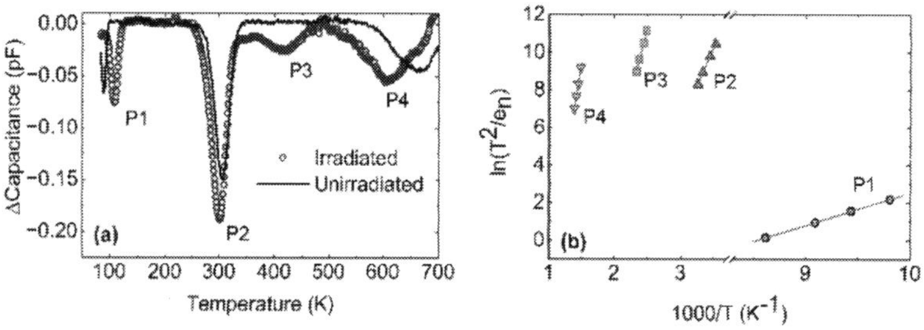

Fig. 5. (e) C-DLTS spectrum obtained for an unirradiated (solid line) the detector irradiated to a neutron fluence of 5 × 10^{11} cm^{-2} (open circles). (f) Arrhenius plots corresponding to the four trap centers observed in the C-DLTS spectra of the irradiated detector.

capacitance mode deep level transient spectra (C-DLTS) of an unirradiated SBD and an irradiated SBD to a neutron fluence of 5 × 10^{11} cm^{-2} obtained using a Sula Technologies DDS-12 deep level transient spectrometer. For the C-DLTS measurements, the diode was biased at a steady-state reverse bias voltage of -12 V. Filling pulses of height 0 V with a width of 2 msec and a period of 5 msec to 1 sec (depending on the rate window setting) were used. Fig. 5(b) shows the Arrhenius plots corresponding to each peak (trap center) from which the activation energies and the capture cross-sections were calculated. Peaks P1and P2 correspond to the Ti(c) impurity and $Z_{1/2}$ centers, which are commonly found in as-grown 4H-SiC [15]. The trap center corresponding to peak P3 is located 1.16 eV below the conduction band minimum (CBM) and is identified as the CAV center [30]. The trap center corresponding to peak P4 is located 1.8 eV below the CBM and could not be identified. Both the trap centers corresponding to P3 and P4 are generally not observed in our as grown 4H-SiC layers and are supposedly created by the neutron irradiation at a fluence of 5×10^{11} cm^{-2}. Although with the increased neutron dosage the optimum bias and the shaping times increased, the CCE decreased, and the effective carrier concentrations decreased to a point transforming the SBD irradiated with the highest fluence to an Ohmic-like detector, the detector was still found to be functional with a moderately high energy resolution of 1.8% for alpha particles incident with 5486 keV energy.

5. Conclusions

The prospect of high-purity and low-defect 4H-SiC epilayers as radiation detectors has been investigated in this article for nuclear reactor dosimetric applications where conventional semiconductor detectors cannot function due to the high radiation background and temperatures. N-type 4H-SiC epitaxial layers with thickness in the range 20 – 250 μm with effective doping concentration on the order of 10^{14} cm^{-3} showed excellent rectification behaviour with very low reverse leakage currents ideal for high resolution radiation detectors. Ni/20μm-4H-SiC SBDs have demonstrated alpha particle detection with excellent energy resolution comparable to that of silicon detectors. Ni/250μm-4H-SiC SBDs irradiated with fast neutrons revealed that the detectors were functional with moderately high resolution even after a neutron fluence of 5×10^{13} cm^{-2}. The effect of the native and radiation-induced defects on the detector performance have been investigated using C-DLTS measurements. The C-DLTS measurements revealed the formation of new types of defects after irradiating the detectors with a neutron fluence of 5×10^{11} cm^{-2}. The capacitance or the junction properties for the higher fluences changed in a way where C-DLTS measurements could no longer be used to characterize the defects. Future work includes alternate defect characterization strategies to detect and identify the defects in these detectors. Also, theoretical calculations are underway to identify the new defects detected in the C-DLTS measurements.

The authors would like to thank Prof. Douglas S. McGregor and Daniel LP Watson of Kansas State University for use of the neutron irradiation facility. One of the authors (KCM) acknowledges the partial financial support obtained from the Department of Energy, Office of Nuclear Energy's Nuclear Energy University Programs (NEUP), Grant No. DE-NE0008662.

References

1. J. Guidez, J. Bodi, K. Mikityuk, E. Giardi, J. Bittan, B. A. Grah, H. Tsige-Tamirat, P. Romojaro, F. Álvarez-Velarde, B. Carluec, ASME J. Nucl. Rad. Sci. **8**, 011312 (2022).
2. N. Almirall, P. B. Wells, S. Pal, P. D. Edmondson, T. Yamamoto, K. Murakami, G. R. Odette, Scr. Mater. **181**, 134 (2020).
3. R. Vuiart, M. Brovchenko, J. Taforeau, V. Jaiswal, E. Drumonteil, Nucl. Sci. Eng. **196**, 455 (2021).
4. S. K. Chaudhuri and K. C. Mandal, "Radiation Detection Using n-Type 4H-SiC Epitaxial Layer Surface Barrier Detectors," in Iniewski K.. (eds) Advanced Materials for Radiation Detection, Cham., Springer, Chapter 9, pp. 183-209 (2022).
5. P. G. Neudeck, D. J. Spry, L. Chen, N. F. Prokop, M. J. Krasowski, IEEE Electron. Dev. Lett. **38**, 1082 (2017).

6. P. G. Neudeck, D. J. Spry, M. J. Krasowski, N. F. Prokop, L. Chen, Mater. Sci. Forum **963**, 813 (2019).
7. K. C. Mandal, J. W. Kleppinger, S. K. Chaudhuri, Micromachines **11**, 254 (2020).
8. F. H. Ruddy, L. Ottaviani, A. Lyoussi, C. Destouches, O. Palais, C. Reynard-Carette, IEEE Trans. Nucl. Sci. **69**, 792 (2022).
9. S. K. Chaudhuri, K. J. Zavalla, K. C. Mandal, Nucl. Instrum. Method Phys. Res. A **728**, 97 (2013).
10. F. H. Ruddy, A. R. Dulloo, J. G. Seidel, M. K. Das, R. Sei-Huang, A. K. Agarwal, IEEE Trans. Nucl. Sci. **53**, 1666 (2006).
11. C. S. Bodie, G. Lioliou, A. M. Barnett, Nucl. Instrum. Meth. Phys. Res. A **985**, 164663 (2021).
12. K. C. Mandal, P. G. Muzykov, S. K. Chaudhuri, J. Russell Terry, IEEE Trans. Nucl. Sci. **60**, 2888 (2013).
13. D. Puglisi and G. Bertuccio, Micromachines **10**, 835 (2019).
14. J. Coutinho, V. J. Torres, I. Capan, T. Brodar, Z. Ereš, R. Bernat, V. Radulović, K. Ambrožič, L. Snoj, Ž. Pastuović, A. Sarbutt, T. Ohshima, Y. Yamazaki, T. Makino, Nucl. Instrum. Meth, Phy. Res. A **986**, 164793 (2021).
15. S. K. Chaudhuri, J. W. Kleppinger, K. C. Mandal, J. Appl. Phys. **128**, 114501 (2020).
16. J. W. Kleppinger, S. K. Chaudhuri, O. Karadavut, K. C. Mandal, Appl. Phys. Lett. **119**, 063502 (2021).
17. J. W. Kleppinger, S. K. Chaudhuri, O. Karadavut, K. C. Mandal, J. Appl. Phys. **129**, 244501 (2021).
18. J. W. Kleppinger, S. K. Chaudhuri, O. F. Karadavut, R. Nag, K. C. Mandal, J. Cryst. Growth **583**, 126532 (2022).
19. J. J. Sumakeris, J. R. Jenny, A. R. Powell, MRS Bull. **30**, 280 (2011).
20. W. Kern, J. Electrochem. Soc. **137**, 1887 (1990).
21. S. K. Chaudhuri, K. J. Zavalla, K. C. Mandal, Appl. Phys. Lett. **102**, 031109 (2013).
22. Shalish and Y. Shapira, IEEE Elctron. Dev. Lett. **21**, 581 (2000).
23. S. M. Sze and K. K. Ng, *Physics of Semiconductor Devices*, (John Wiley & Sons, New Jersey, 2007).
24. R. T. Tung, Appl. Phys. Rev. **1**, 011304 (2014).
25. S. Seshadri, A. R. Dulloo, F. H. Ruddy, J. G. Seidel, L. B. Rowland, IEEE Trans. Electron Dev. **46**, 567 (1999).
26. M. B. H. Breese, J. Appl. Phys. **74**, 3789 (1993).
27. J. F. Ziegler, M. D. Ziegler, J. P. Biersack, Nucl. Instrum. Meth. Phys. Res. B **268**, 11, 1818 (2010).
28. K. C. Mandal, S. K. Chaudhuri, K. V. Nguyen, M. A. Mannan, IEEE Trans. Nucl. Sci. **61**, 2338 (2014).
29. J. W. Kleppinger, S. K. Chaudhuri, O. Karadavut, R. Nag, D. L. Watson, D. S. McGregor, K. C. Mandal, IEEE Trans. Nucl. Sci. Early Access, Apr. 2022.
30. N. T. Son, P. Stenberg, V. Jokubavicius, H. Abe, T. Ohshima, J. U. Hassan, I. G. Ivanov, Appl. Phys. Lett. **114**, 212105 (2019).

Fast neutron fluence estimation by measurement of ^{93m}Nb and ^{93}Mo

Masashi Ichikawa[1,*], *Shigeaki* Tanaka[2], and *Kazuhiro* Chatani[2]

[1] Nippon Nuclear Fuel Development Co., Ltd., 2163, Narita-Cho, Oarai-Machi, Higashi-Ibaraki-Gun, Ibaraki-Ken, 311-1313, Japan
[2] Toshiba Energy Systems & Solutions Corp., 8, Shinsugita-Cho, Isogo-Ku, Yokohama-Shi, 235-8523, Japan

Abstract. The possibility for estimation of fast neutron fluence from ^{93m}Nb produced from Nb contained in structural component materials of nuclear power plants was investigated as an evaluation tool for extending plant operating lifetime. First, to establish a measurement method for ^{93m}Nb in structural component materials, chemical separations and measurements of ^{93m}Nb were carried out using specimens of fast neutron-irradiated XM-19 and X-750 alloy; these are used as structural component materials and contain Nb as a component. The ^{93m}Nb activities obtained from the measurements of both XM-19 and X-750 alloy agreed well with the calculated values of ^{93m}Nb, which were simply calculated from the evaluated fast neutron fluence. On the other hand, in the case of type 316L stainless steel, which contains Mo, Mo derived ^{93m}Nb presents a problem, so an approach was used that estimates the amount of Mo derived ^{93m}Nb based on ^{93}Mo measurements with the third specimen, neutron-irradiated type 316L stainless steel. The difference between the measured and calculated ^{93m}Nb values was about two-fold, indicating that the estimation method for ^{93m}Nb produced from Mo needs to be improved.

1 Introduction

One of the approaches for improving the estimation accuracy of fast neutron fluence experienced by nuclear power plant structural component materials is to estimate the fast neutron fluence based on the measurement of the activity of the activated nuclides in the materials. This means that, the results of measurements made after chemical separations of nuclides produced by the reaction between elements contained in the structural materials and fast neutrons are reflected in the estimation of the fast neutron fluence. ^{93m}Nb produced by the ^{93}Nb(n, n')^{93m}Nb reaction between Nb and fast neutrons has a relatively long half-life of 16.13 years and is used to estimate long-term fast neutron fluence. Types 304 and 316L stainless steels (SSs) are mainly used for structural components such as core shrouds and upper grid plates in BWRs, and it has been reported that type 304 SS contains several ppm to several tens of ppm of Nb [1], which may allow estimation of fast neutron fluence by the ^{93}Nb(n, n')^{93m}Nb reaction. In fact, estimations of fast neutron fluence from the ^{93}Nb(n,

* Corresponding author: ichikawa@nfd.co.jp

n')^{93m}Nb reaction have been made using Nb present in structural components on the order of several tens of mass ppm [2-4]. On the other hand, in the case of a material containing Mo such as type 316L SS, ^{93m}Nb is also produced from Mo in that structural material [3, 4], so the amount of ^{93m}Nb derived from Mo must be estimated. ^{93m}Nb from Mo is produced by the electron capture decay of ^{93}Mo that was produced from the thermal neutron capture reaction of ^{92}Mo, which is one of the stable isotopes of Mo. Therefore, in the case of a material containing Mo, ^{93m}Nb produced from both Mo and Nb is included, and if the activity of ^{93m}Nb produced from Mo is not negligible, a correction is necessary to subtract the activity of ^{93m}Nb produced from Mo from the ^{93m}Nb measured activity. (Of course, if the activity of ^{93m}Nb produced from Mo is overwhelmingly greater than the activity of ^{93m}Nb produced from Nb, the fast neutron flux from ^{93}Nb(n, n')^{93m}Nb cannot be estimated.) As a method for accurately estimating the produced amount of ^{93m}Nb from Mo, an estimation from the measured value of ^{93}Mo is effective, as ^{93}Mo is an intermediate product in the production of ^{93m}Nb from Mo.

In this study, the ^{93m}Nb measurement method was established using irradiated specimens of XM-19 and X-750 alloy containing Nb as a constituent element whose amount is within the material specifications such as set by the ASTM. The validity of the measurement method was verified by comparing the measured value with the calculated value of ^{93m}Nb estimated from the evaluated fast neutron fluence. The validated ^{93m}Nb measurement method was used to measure ^{93m}Nb for the third specimen, an irradiated type 316L SS specimen. To estimate the amount of ^{93m}Nb produced from Mo, ^{93}Mo was also measured for the above three specimens and the amount of ^{93m}Nb produced from Mo was estimated. The activity of ^{93m}Nb and ^{93}Mo of each specimen were measured by X-ray measurements with a Ge semiconductor detector after dissolving each specimen and separating ^{93m}Nb and ^{93}Mo from each other by an ion exchange separation method. The chemical separation between ^{93m}Nb and ^{93}Mo is necessary, because both are measured by X-ray measurements and both are measured at the same X-ray energy of 16.6 keV.

2 Experimental

2.1 Specimens

Irradiated XM-19, X-750 alloy and type 316L SS, which are used as structural component materials, were used as specimens and there was one test piece for each. XM-19 and X-750 alloy were used as specimens containing Nb as a component to establish the ^{93m}Nb measurement method. Table 1 summarizes their compositions listed on the respective mill test reports. In the absence of Nb or Mo concentration data on the mill test reports, the concentration was determined by measuring the solution of each specimen by inductively coupled plasma mass spectrometry (ICP-MS). The concentration measurements were performed after confirming sufficient accuracy in advance.

All specimens were irradiated in the Japan Materials Testing Reactor (JMTR) under the conditions shown in Table 2. The fast neutron fluence integrated by the specimens is equivalent to the cumulative irradiation in the core shrouds of Japanese BWRs during a period of 60 years.

Table 1. Composition of specimens.

Material	Composition (wt%)											
	C	Si	Mn	P	S	Ni	Cr	Mo	Nb	Ti	Al	Fe
XM-19	0.03	0.44	5.23	0.007	0.002	12.14	21.23	2.20	0.21	0.004	0.01	Bal.
X-750 alloy	0.047	0.13	0.11	0.007	0.003	72.18	15.42	—	0.99	2.68	0.60	7.6
316L SS	0.008	0.51	0.95	0.018	0.006	12.58	16.49	2.16	—	—	—	Bal.

Table 2. Irradiation conditions of specimens.

Reactor	JMTR (Thermal Power: 50MW)
Period	Jan. 1, 1999-May 14, 2001
Cycle	12 cycles
Time	299 days (2.58×10^{7} s)
Max. neutron fluence (Fast)	2.66×10^{21} n/cm^2 (E > 1 MeV)
Neutron fluence (Thermal)	4.13×10^{21} n/cm^2 (E < 0.683 eV) at max. neutron fluence (Fast)
Neutron flux (Fast)	1.03×10^{14} n/cm^2 s (E > 1 MeV)
Neutron flux (Thermal)	1.60×10^{14} n/cm^2 s (E < 0.683 eV) at max. neutron fluence (Fast)

2.2 Measurements

2.2.1 Separation of Mo and Nb

The separation of Nb and Mo was performed by applying an ion exchange separation method using a hydrofluoric acid system, with reference to the ^{93m}Nb separation method of Serén and Kekki [4] and the report of Fujimoto and Shimura [5] about element adsorption behaviors on an anion exchange resin in hydrofluoric acid and hydrochloric acid systems.

Approximately 20 mg of each specimen was dissolved in a mixture of nitric acid (HNO_3), hydrochloric acid (HCl), and hydrofluoric acid (HF) while heating on a hot plate. Then, the acidic solutions of dissolved specimens were evaporated to dryness by further hot plate heating. The residues were dissolved in 2 mol/L HF, and passed through an anion exchange column filled with DOWEX 1X-8 200-400 mesh. In the next step, 2 mol/L HF was passed through the column. This step flushed out the main structural material elements such as Fe, and the major radionuclides such as ^{60}Co, from the column. Next, Mo was eluted from the column using a mixture of 8 mol/L HF and 4 mol/L HCl. Finally, Nb was eluted from the column using 1 mol/L HCl.

2.2.2 Measurement of ^{93}Mo

The eluant of 8 mol/L HF and 4 mol/L HCl containing Mo was evaporated to dryness by heating on a hot plate and the residue was dissolved in 1 mol/L HNO_3. Then, 1 mg of lanthanum was added to the solution, and lanthanum hydroxide precipitate was formed by adding ammonia water. In this case, Mo remained in the liquid phase, while ^{60}Co, which could not be removed by ion exchange separation, was included on the precipitate side. After a vacuum filtration, 100 μg of Mo was added to the collected filtrate, which was then adjusted to a slightly acidic condition as confirmed by pH paper testing by adding HCl. Then bromine water and an ethanolic solution of α-benzoin oxime were added to get Mo precipitate. The precipitate was collected on a membrane filter (47 mm diameter; 0.45 μm pore size). The activity of ^{93}Mo in the precipitate was determined with a high-purity germanium low energy photon spectrometer (LEPS) by measuring the 16.6 keV radiation emitted by ^{93}Mo. To estimate the recovery, the amount of Mo in the precipitate was determined by X-ray fluorescence (XRF). The LEPS and XRF apparatus were calibrated using standard samples.

2.2.3 Measurement of ^{93m}Nb

To remove HF, the 1 mol/L HCl solution containing Nb was evaporated to dryness by heating on a hot plate and the residue was dissolved with 1 mol/L HCl. After 500 μg of lanthanum was added to the solution, ammonia water was added to precipitate lanthanum hydroxide, and Nb was coprecipitated. The precipitate was collected using a membrane filter (47 mm diameter; 0.45 μm pore size). The activity of ^{93m}Nb in the precipitate was determined with the LEPS by measuring the 16.6 keV radiation emitted by ^{93m}Nb. To estimate the recovery of Nb, the activity of ^{94}Nb in the precipitate was determined with a germanium semiconductor detector. The LEPS and germanium semiconductor detectors were calibrated using standard samples.

2.3 Calculations

2.3.1 Activity of ^{93m}Nb from ^{93}Nb

The activity of ^{93m}Nb produced from ^{93}Nb ($A_{Nb93m(Nb)}$) is calculated as follows:

$$A_{\mathrm{Nb93m(Nb)}} = N_{\mathrm{Nb93}}\,\sigma_{\mathrm{Nb93}}\,\varphi_{\mathrm{f}}\left(1 - e^{-\lambda_{\mathrm{Nb93m}}\,t}\right)e^{-\lambda_{\mathrm{Nb93m}}\,t_m} \quad (1)$$

where, N_{Nb93} is the number of atoms of ^{93}Nb in the specimen, σ_{Nb93} is the cross section of the reaction ^{93}Nb(n, n')^{93m}Nb, φ_f is the fast neutron flux, λ_{Nb93m} is the decay constant of ^{93m}Nb, t is the irradiation time, and t_m is the time from the end of irradiation to the measurement.

In the calculations of this study, the cross section of ^{93}Nb(n, n')^{93m}Nb and the fast neutron flux were assumed to be constant. For the fast neutron flux values, the values shown in Table 2 were used. For the cross section values, the values corresponding to 2 MeV, the average energy of the fission spectrum, were used. The activity of ^{93m}Nb calculated from equation (1) was compared with the measured activity of ^{93m}Nb for each specimen to confirm the validity of the measured values.

2.3.2 Activity of ^{93m}Nb from ^{93}Mo

^{92}Mo produces ^{93}Mo by thermal neutron capture, and ^{93}Mo disintegrates to ^{93m}Nb. In this study, it was assumed that all the ^{92}Mo reacted with neutrons to become ^{93}Mo, and all the ^{93}Mo formed decayed to ^{93m}Nb. The activity of ^{93m}Nb produced from ^{93}Mo ($A_{Nb93m(Mo)}$) is calculated as follows:

$$A_{\mathrm{Nb93m(Mo)}} = \lambda_{\mathrm{Nb93m}}\,N'_{\mathrm{Nb93m}} \quad (2)$$

$$N'_{\mathrm{Nb93m}} = \frac{\lambda_{\mathrm{Mo93}}}{\lambda_{\mathrm{Nb93m}} - \lambda_{\mathrm{Mo93}}}\,N_{\mathrm{Mo93}}\left(e^{-\lambda_{\mathrm{Mo93}}\,t_m} - e^{-\lambda_{\mathrm{Nb93m}}\,t_m}\right) + N_{\mathrm{Nb93m}}\,e^{-\lambda_{\mathrm{Nb93m}}\,t_m} \quad (3)$$

where, N'_{Nb93m} is the number of atoms of ^{93m}Nb produced from ^{92}Mo during the period between irradiation and measurement, λ_{Mo93} is the decay constant of ^{93}Mo, and N_{Mo93} and N_{Nb93m} are the numbers of ^{93}Mo and ^{93m}Nb atoms at the end of irradiation. N_{Mo93} and N_{Nb93m} are calculated as follows:

$$N_{Mo93} = N_{Mo92}\, \sigma_{Mo92}\, \varphi_{th} \left(\frac{e^{-\sigma_{Mo92}\varphi_{th} t}}{\lambda_{Mo93} - \sigma_{Mo92}\varphi_{th}} + \frac{e^{-\lambda_{Mo93} t}}{\sigma_{Mo92}\varphi_{th} - \lambda_{Mo93}} \right) \quad (4)$$

$$N_{Nb93m} = N_{Mo92}\, \sigma_{Mo92}\, \varphi_{th}\, \lambda_{Mo93} \left[\frac{e^{-\sigma_{Mo92}\varphi_{th} t}}{(\lambda_{Mo93} - \sigma_{Mo92}\varphi_{th})(\lambda_{Nb93m} - \sigma_{Mo92}\varphi_{th})} + \frac{e^{-\lambda_{Mo93} t}}{(\sigma_{Mo92}\varphi_{th} - \lambda_{Mo93})(\lambda_{Nb93m} - \lambda_{Mo93})} + \frac{e^{-\lambda_{Nb93m} t}}{(\sigma_{Mo92}\varphi_{th} - \lambda_{Nb93m})(\lambda_{Mo93} - \lambda_{Nb93m})} \right] \quad (5)$$

where, N_{Mo92} is the initial number of atoms of ^{92}Mo in the specimen, σ_{Mo92} is the cross section of ^{92}Mo(n, γ)^{93}Mo, and φ_{th} is the thermal neutron flux. Here, φ_{th} is the thermal neutron flux that contributes to the reaction of ^{92}Mo(n, γ)^{93}Mo, and it was estimated from the measured activity of ^{93}Mo along with σ_{Mo92}. N_{Mo93} in equation (4) can be estimated from equation (6) based on the measured activity of ^{93}Mo (A_{Mo93}).

$$N_{Mo93} = \frac{A_{Mo93}}{\lambda_{Mo93}\, e^{-\lambda_{Mo93} t_m}} \quad (6)$$

When N_{Mo93} is obtained, $\sigma_{Mo92}\,\varphi_{th}$ can also be obtained from equation (4). Furthermore, once $\sigma_{Mo92}\,\varphi_{th}$ is obtained, N_{Nb93m} can also be obtained from equation (5).

Therefore, the procedure to obtain the activity of ^{93m}Nb produced from Mo based on the measured activity of ^{93}Mo is the following.

First, the number of atoms of ^{93}Mo at the end of irradiation, N_{Mo93}, is obtained from the measured activity of ^{93}Mo by equation (6). Next, after substituting the obtained N_{Mo93} into equation (4), the product of the cross section and thermal neutron flux, $\sigma_{Mo92}\,\varphi_{th}$, is obtained. Then, by substituting $\sigma_{Mo92}\,\varphi_{th}$ into equation (5), the number of atoms of ^{93m}Nb produced from ^{92}Mo at the end of irradiation, N_{Nb93m}, is obtained. By substituting N_{Mo93} and N_{Nb93m} into equation (3), the number of atoms of ^{93m}Nb produced from ^{92}Mo during the period between irradiation and measurement, N'_{Nb93m}, is obtained. Finally, N'_{Nb93m} is converted to the activity of ^{93m}Nb, $A_{Nb93m(Mo)}$, by equation (2).

In this study, the activity of ^{93m}Nb produced from Mo in each specimen was evaluated using the measured activity of ^{93}Mo in each specimen and the above procedure.

3 Results and discussion

Table 3 lists the concentrations of Nb and Mo and the activities of ^{93m}Nb and ^{93}Mo for each specimen. The Nb/Mo ratio shown in Table 3 indicates the concentration ratio of Nb and Mo in each specimen. The Nb concentration of type 316L SS and the Mo concentration of X-750 alloy are the values obtained by measuring the solution of each specimen by ICP-MS. The Nb concentration of type 316L SS is 0.00054 % (= 5.4 ppm), which is similar to the reported analytical value of type 304 SS [1].

Table 3. Results of analysis.

Material	Mass (g)	Nb concentration in specimen (wt%)	Mo concentration in specimen (wt%)	Nb/Mo ratio	^{93m}Nb activity (Bq)	^{93}Mo activity (Bq)
XM-19	0.0192	0.21	2.2	9.5×10^{-2}	8.9×10^{4}	1.2×10^{2}
X-750 alloy	0.0196	0.99	0.27	3.7	4.7×10^{5}	1.3×10^{1}
316L SS	0.0200	0.00054	2.16	2.5×10^{-4}	6.2×10^{2}	1.6×10^{2}

Table 4 allows a comparison of the measured activity of ^{93m}Nb (M) and the calculated activity of ^{93m}Nb produced from Nb (C). The ratios of the measured and calculated values of XM-19 and X-750 alloy are 1.1 and 1.2, which are in relatively good agreement, and the measured values of ^{93m}Nb obtained by the measurement method in this study are reasonable.

On the other hand, in type 316L SS, the ratio of measured to calculated values is 2.8, and the agreement is poor. Focusing on the Nb/Mo ratio in each specimen, the ratio of type 316L SS is lower than the ratios of the other two. Baers and Hasanen [3] reported that the Nb/Mo ratio must be greater than a certain value (Nb/Mo ratio > 0.01 in their study) for the ^{93m}Nb produced from Nb to be dominant. This means that when the Nb/Mo ratio is low, the proportion of ^{93m}Nb produced from Mo in the activity of ^{93m}Nb in the specimen becomes large, and ^{93m}Nb produced from Mo cannot be neglected. In type 316L SS, the Nb/Mo ratio is low and the measured activity is higher than the calculated activity. This difference between measured and calculated activities suggests that the actual activity (i.e., the measured activity) was affected by ^{93m}Nb produced from Mo, resulting in a discrepancy between the measured and calculated values.

Table 5 shows the results of estimating the activity of ^{93m}Nb produced from ^{92}Mo based on the measured activity of ^{93}Mo shown in Table 3. The calculated activity of ^{93m}Nb produced from Mo in XM-19 and X-750 alloy is sufficiently small compared to the measured activity of ^{93m}Nb as shown in Table 4. On the other hand, in type 316L SS, the calculated activity of ^{93m}Nb produced from Mo is only about one order of magnitude smaller than the measured value, which is a non-negligible amount.

Table 6 is a comparison of the measured activity of ^{93m}Nb after subtracting the activity of ^{93m}Nb produced from ^{92}Mo (M') and the calculated activity of ^{93m}Nb produced from Nb (C). Even after subtracting the activity of ^{93m}Nb produced from Mo, the ratio of measured to calculated values in type 316L SS is 2.4, which is a slight improvement, but there is still a discrepancy compared to the XM-19 and X-750 alloy values. The reason for the discrepancy between the calculated and measured values may be due to uncertainty in the measured value of ^{93}Mo, uncertainty in the calculation model for the amount of ^{93m}Nb produced from ^{92}Mo, or other factors. The calculation of ^{93m}Nb produced from ^{92}Mo should be done in detail using an analysis code such as ORIGEN2[6].

Table 4. Comparison of the measured activity of ^{93m}Nb and the calculated activity of ^{93m}Nb produced from Nb.

Material	M Measured activity of ^{93m}Nb (Bq)	C Calculated activity of ^{93m}Nb produced from Nb (Bq)	M/C
XM-19	8.9×10^{4}	8.4×10^{4}	1.1
X-750 alloy	4.7×10^{5}	4.0×10^{5}	1.2
316L SS	6.2×10^{2}	2.2×10^{2}	2.8

Table 5. Calculated activity of ^{93m}Nb produced from ^{92}Mo.

Material	Calculated activity of ^{93m}Nb produced from ^{92}Mo (Bq)
XM-19	6.5×10^1
X-750 alloy	7.1
316L SS	8.7×10^1

Table 6. Comparison of the measured activity of ^{93m}Nb subtracting ^{93m}Nb produced from ^{92}Mo and the calculated activity of ^{93m}Nb produced from Nb.

Material	M' Measured activity of ^{93m}Nb subtracting ^{93m}Nb produced from ^{92}Mo (Bq)	C Calculated activity of ^{93m}Nb produced from Nb (Bq)	M'/C
XM-19	8.9×10^4	8.4×10^4	1.1
X-750 alloy	4.7×10^5	4.0×10^5	1.2
316L SS	5.3×10^2	2.2×10^2	2.4

4 Conclusions

In this study, the measurement method of ^{93m}Nb was confirmed using three irradiation specimens of XM-19, X-750 alloy and type 316L SS. In addition, as an estimation method of Mo derived ^{93m}Nb when the influence of ^{93m}Nb produced from Mo cannot be ignored, an estimation method of ^{93m}Nb produced from Mo based on the measured ^{93}Mo was examined.

The measured activity of ^{93m}Nb for XM-19 and X-750 alloy agreed well with the calculated activity assuming that ^{93}Nb is the source, and it was confirmed that the method used in this study can obtain reasonable ^{93m}Nb measurements.

On the other hand, in type 316 L SS with low Nb concentration relative to Mo concentration, compared to XM-19 and X-750 alloy, the measured activity of ^{93m}Nb did not agree well with the calculated activity assuming that ^{93}Nb is the source, and the effect of ^{93m}Nb from Mo was confirmed. Therefore, ^{93m}Nb from Mo was evaluated based on ^{93}Mo measurements and subtracted from the ^{93m}Nb measurements. However, the difference between the measured and calculated values of ^{93m}Nb was still about two-fold, indicating that the estimation method for ^{93m}Nb produced from Mo needs to be improved. It is planned to investigate the amount of ^{93m}Nb produced from Mo in detail in the future.

In addition, type 304 SS, which was not treated in this study, does not contain Mo as a constituent element, and there is a possibility that ^{93m}Nb produced from Mo can be neglected, and it is planned to study the feasibility of fast neutron fluence estimation from Nb contained in type 304 SS also.

References

1. N. Kawata, Y. Shiratori, K. Maekawa, O. Arai, M. Shimizu, "Analysis of Elemental Composition of Main Construction Materials in the Nuclear Power Plant "Fugen"", J. At. Energy Soc. Jpn., **9**, 405-418 (2010) (In Japanese)
2. J. van Aarle, I. Guenther, F. Hegedues, F. Gabler, "Retrospective Fast Neutron Dosimetry of Nuclear Power Plants by Means of Scraping Samples Using the

^{93}Nb(n,n')^{93m}Nb Reaction", in: *Reactor Dosimetry: Radiation Metrology and Assessment*, Williams, J.G. et al (eds.), (ASTM International, West Conshohocken, PA, 2001)

3. L.B. Baers, E.K. Hasanen, ASTM Special Technical Publication, 1228, 205-214 (1994)
4. T. Serén, T. Kekki, Retrospective dosimetry based on niobium extraction and counting – VTT's contribution to the RETROSPEC project (2003); available at https://www.vttresearch.com/sites/default/files/pdf/tiedotteet/2003/T2203.pdf
5. K. Fujimoto, M. Shimura, "Determination of trace amounts of elements in high-purity ion by ICP-MS after ion chromatographic separation using the cumulated bed of cation- and anion-exchange resin", Bunseki Kagaku, **50**, 175-182 (2001) (In Japanese)
6. A. G. Croff, "ORIGEN2: A Versatile Computer Code for Calculating the Nuclide Compositions and Characteristics of Nuclear Materials", Nuclear Technology, 62, 335-352 (1983).

Product names mentioned herein may be trademarks of their respective companies/organizations.

Dosimetry data from fuel channel clips for benchmarking a new computational fluid dynamics model in neutron transport

Michael Manahan[1,*]

[1]MP Machinery and Testing (MPM), LLC, 2161 Sandy Drive, State College, PA 16803, USA

Abstract. Accurate neutron transport models for BWRs are needed to characterize neutron damage to the top guide, top guide cylinder, vessel nozzles, and to other upper internals. In the past, neutron transport calculations above top-of-active-fuel (TAF) have had large uncertainties (well in excess of ± 20%) mainly due to the fact that the steam density distribution in that region is not well known. An advanced three-dimensional (3D) neutron transport model, which incorporates computational fluid dynamics (CFD) in the code suite, has been developed. Retrospective dosimetry measurements were made to benchmark the transport results. Six fuel channel clips were removed from the top of the fuel bundles, and small disk-shaped dosimetry samples were cut from the clips. The cuts were made through the Inconel spring and through the stainless steel clip body. The clips were selected from bundles that cover regions of low, intermediate, and high steam density. The average C/M ratio for the 10 dosimeters is 1.06. It has been shown that all of the calculated dosimeter activities fall within ± 20% of the measurement. This meets the criterion set by RG 1.190 [1] for acceptability of the calculations.

1 Introduction

Previous neutron transport calculations above the top-of-active-fuel (TAF) in BWRs have had large uncertainties mainly due to the fact that the water density distribution in that region was not well known. Estimates of the uncertainties for fluences to the top guide and upper vessel regions are ~50% and higher. Previous neutron transport calculations have assumed a constant steam void fraction, typically ~80 % void fraction, in the region between TAF and the steam separator inlets. Although it was recognized that this is a coarse approximation, an accurate spatial distribution of the steam density was not available in the past. As a result of the approximation, the analytic uncertainty analysis of fluences to components located above the grid beam resulted in values far in excess of 20 percent. Regulatory Guide 1.190 targets uncertainties to be within 20% for the beltline region calculations, and it is reasonable to assume that this level of uncertainty can be achieved for fluences calculated outside of the active fuel region.

* Corresponding author: MPManahan@MPMachineryandTesting.com

2 Computational fluid dynamics (CFD) model

Fluid flow through a BWR presents a complex three-dimensional fluid dynamics problem due to the presence of multiple fluids interacting with each other. The fluid flow and mixing is governed by factors such as the physical properties of these fluids, flow velocities, geometry of the flow channel, and the interphase heat and mass transfer between fluids. The flow that leaves the core consists of steam and water, and these two components are partitioned as follows: steam flow from the bundles; subcooled water-rod flow; and subcooled bypass flow in-between the bundles. The water-rod flow mixes with the bundle steam near the top of the bundles below the upper tie plate (UTP). As this mixed fluid and the bypass water emerge from the bundles, they mix with each other in the space above the top guide (also referred to as the grid beam). This fluid mixing and flow evolution process dictates the properties of the composite fluid in the region between TAF and the steam separator dome.

2.1 CFD model description

The overall objective of the CFD modelling effort is the determination of the spatial distribution of the volume fraction of steam in the region above TAF. To accomplish this, a fluid volume mesh was created that extends from the grid beam top surface to the spherical dome-shaped surface at which the steam enters the stand-pipes in the steam separator dome. As with the TORT [2] neutron transport model, it is only necessary to create a quarter-core CFD model due to geometric symmetry. The 0° to 90° quadrant was modeled for the three plant (operating US BWR) cycles for which the dosimeters were analyzed. Figure 1 is a 3D view of the quarter-symmetric model showing the top guide cylinder, grid beam, steam separator, spargers, and fuel bundles in their respective positions. Outside of the shroud, the downcomer water is subcooled and is modeled as such.

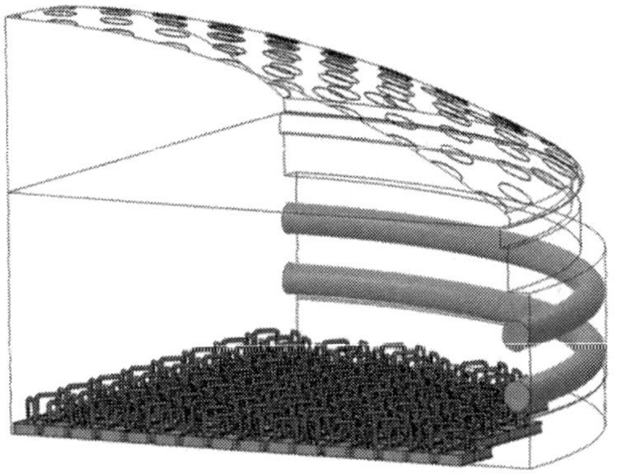

Fig. 1. 3D view of CFD model showing steam separator, top guide, grid beam, spargers, and fuel bundles.

The three fuel cycles analyzed as part of the benchmarking have two or more fuel types. In order to capture the physics for the three-dimensional flow accurately, the CFD geometry included models of all of the fuel types at their respective positions in the core. Further, each fuel bundle is accompanied by a bail-handle and an UTP which acts as a flow obstruction for the steam exiting the fuel bundle. Both the UTP and fuel rods were modelled so that the inlet boundary conditions provided by the SIMULATE-3 [3] output could be accurately represented.

2.2 CFD model output

An analysis was conducted to determine whether a cycle-by-cycle CFD analysis is needed in order to improve the neutron transport model above TAF to yield accurate calculated-to-measured (C/M) ratios. It was concluded that a cycle-by-cycle CFD analysis is needed. Not only are there large average density differences between cycles, but there are also large differences in the spatial distribution of steam for each cycle. The observed cycle differences are due mainly to differences in fuel design. An important fuel design difference is the size of the gap between bundles. This gap between bundles is related to the outer channel dimensions and the bundle corner geometries. A larger gap between bundles, and between bundles and the grid beam, allows for a larger flow area for the subcooled bypass water. This affects the spatial distribution in the mixing region above the grid beam. Also, flow areas for the bundles affect flow velocities. Therefore, bundle geometry plays an important role in the spatial distribution of the steam. The void fraction inlet conditions play a critical role in both the overall average fluid density and the fluid density distribution. Because void fraction has a direct effect on fluid density, the average void fraction of the bundles affects the overall fluid density. Further, the bundle locations affect the density distribution across the fluid volume. Bundle flow rates also have an effect on the steam density distribution.

An example water volume fraction distribution is shown in Figure 2 along with the radial and azimuthal distribution at an elevation of 50 cm above the grid beam. The previous assumption of an 80% void fluid inside the top guide and above the grid beam was found to be overly conservative by a significant amount over most of the steam dome region. The average bundle exit void fraction across the bundles for each cycle analyzed ranged from 74.2% to 80.5%, making 80% void seem like a good assumption. However, the CFD simulations show that while the void fraction for the steam above the grid beam is less than 80% over the central core region, much higher density water is observed near the periphery which causes large reductions in neutron flux at the top guide and upper vessel region. The lower density fluid in the central core region reduces the likelihood of neutron scattering toward the top guide, but this effect is small compared to the reduction associated with high density steam at the periphery.

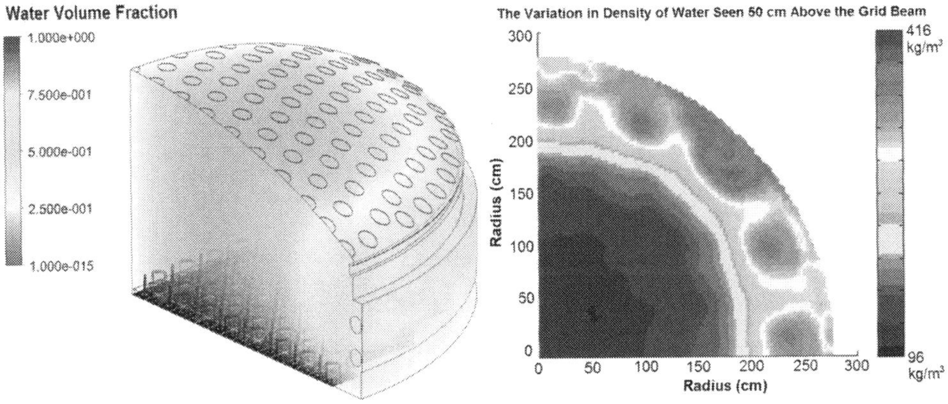

Fig. 2. (Left) Example of CFD determined water volume fraction distribution. High density steam is observed between the top guide and the outer fuel bundles. (Right) Plot showing the radial and azimuthal variation of water density at an elevation 50 cm above the grid beam.

2.3 Effect of water density distribution on neutron transport

As discussed, the steam density in the central portion of the top guide volume is lower than previously assumed in neutron transport models, and the density at the periphery near the top guide cylinder inner diameter surface is much higher. TORT analyses were conducted to quantify the impact of the CFD results. A large reduction in the flux is observed for upper region top guide cylinder welds and also for the pressure vessel. An example of the flux reduction to top guide cylinder vertical welds V5 and V6 is shown in Figure 3. The reduction in flux is a result of high density steam which is observed in the CFD output between the top guide and the outer fuel bundles (Figure 2). This steam acts as a very effective shield for the top guide cylinder and upper vessel, reducing the flux to approximately 1/6th of the previous 80% void case transport fluxes.

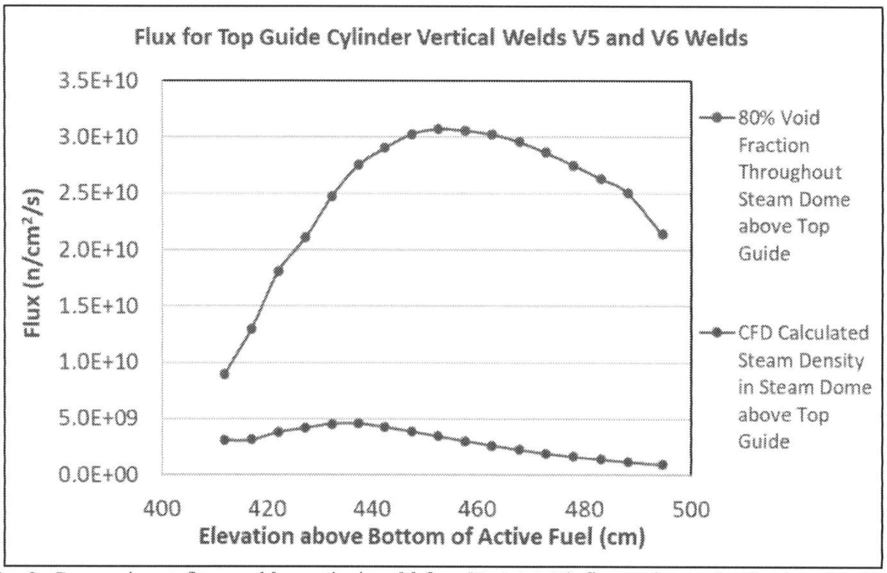

Fig. 3. Comparison of top guide vertical weld fast (E>1 MeV) fluxes determined using the CFD calculated water density compared to the fast flux determined assuming 80 % void fraction above the grid beam.

Similarly, Figure 4 shows the peak vessel fast (E>1 MeV) fluence at 54 EFPY calculated using both the CFD water density distribution and the 80% void fraction assumption. Of particular interest is the non-physical peak shown in Figure 4 that occurs above the grid beam elevation (>400 cm elevation) in the 80% void model. This erroneous peak is due to the direct shine from the top of the core in the 80% void assumption case. The peak is not physically correct because the high density water between the core edge and the top guide cylinder provides very effective shielding which suppresses this local peak and keeps the fluences below 1.0 E17 n/cm^2 in this region of the vessel. If the peak were real, it may have caused several issues for the plant since it would require the inclusion of neutron damage effects for the upper shell course materials in the PT curves. In such a case, the upper shell course materials would need to be considered in the plant pressure vessel surveillance program.

3 CFD benchmark results

Conventional dosimetry was not available for benchmarking the CFD model. Therefore, a decision was made to cut material from fuel channel fasteners for this purpose. Six fuel channel fasteners (referred to as fuel clips) were removed from bundles which were irradiated in cycles leading up to the clip removal (cycles 19, 20, and 21). Two of the clips were irradiated in all three cycle (GEQ771/GEQ846), two were in C20 and C21 (GER212/GER238), and two were in C21 only (GES366/GES547). Other than one bundle (GER212), all of the bundles used for dosimetry were shuffled to a new location each cycle. As a result, the clips were exposed to widely varying water densities, thereby providing a very good benchmarking opportunity. Analysis of clip GER212, that was introduced in C20, was not done because the adjacent control rod was inserted about 60% of the way through C21 due to a failed fuel issue.

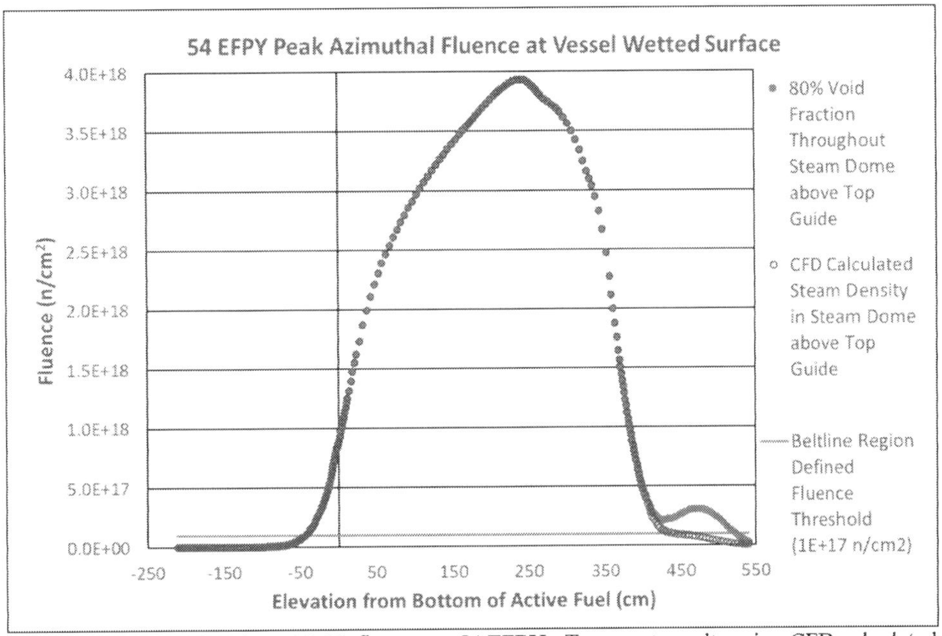

Fig. 4. Peak vessel fast (E>1 MeV) fluence at 54 EFPY. Transport results using CFD calculated water density are compared with the results assuming 80 % void fraction above the grid beam.

After the conclusion of cycle 21, the 6 clips were removed from the bundles and stored in a holder in the spent fuel pool. As a result of the high dose rates, MPM designed and built special cutting equipment to extract small disks from the clips for the specific activity measurements. A fuel channel clip after cutting is shown in Figure 5. The clips consist of an outer Inconel spring which is attached to the stainless steel body via a captive screw. Since the cuts were made through both the spring and fastener body, two dosimeter disks were obtained from each clip.

The primary focus of the CFD benchmark analysis is on the Fe54(n,p)Mn54 reaction. The short half-life of 70.82 days for the Ni58(n,p)Co58 reaction precluded obtaining accurate counting results. The only other reaction that could be considered is the Nb-93(n,n')Nb-93m reaction, but the high activity of the dosimetry materials precluded accurate counting for this reaction.

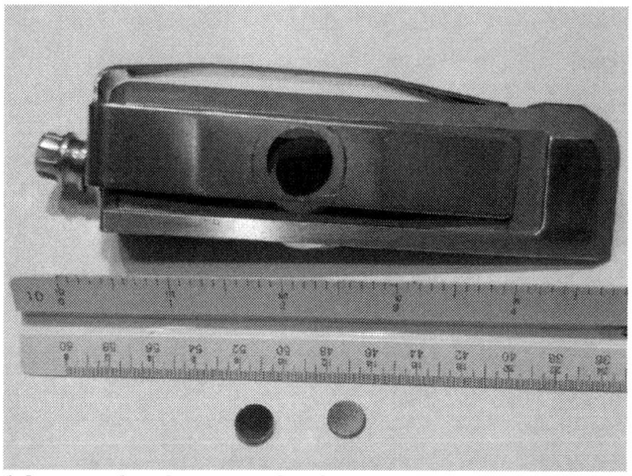

Fig. 5. Fuel channel fastener showing example of cut made to extract the retrospective dosimetry disks for specific activity measurements.

As mentioned, the TORT transport models with the CFD water density inputs were run for cycles C19 through C21, and a through-cycle analysis was included in the C21 calculations. The calculated fluences were determined at the center of the cut location for each clip. The accumulated fluence calculations included the fact that two of the bundles were moved to different locations at the end of C19 and C20, and two other bundles were moved at the end of C20. The calculation accounted for the increase in the channel length from 32.84 cm to 33.60 cm during operation due to channel swelling. The swelling model correlates channel length increase as a function of exposure. So, the channel growth axial span indicated in Figure 6 is largest for the bundles that were in for 3 cycles, and the lower limit is for the bundles that were introduced in C21. As shown in the figure, the clip sampling location is just above the top surface of the grid beam where the channel exit steam mixes with the subcooled bypass water. This is a challenging elevation for fluence determination as the water density in localized regions varies from 80 to 720 kg/m^3 across the core.

The Inconel and SS disks were lightly abraded and cleaned to remove oxide from the samples prior to counting and weighing. Radiometric analysis was performed using high resolution gamma emission spectroscopy. The total mass of each disk was measured using a high precision analytical digital balance. In retrospective dosimetry applications, it is necessary to measure the composition of the alloy. X-ray fluorescence measurements were made to determine the iron weight percent for each of the 10 disks. All iron mass measurements were made after irradiation.

The CFD model benchmark results are summarized in Table 1. The average C/M ratio for the 10 dosimeters is 1.06. For each dosimetry result, it has been shown that all of the calculated dosimeter activities fall within ± 20% of the measurement. This meets the criterion set by RG 1.190 for acceptability of the calculations.

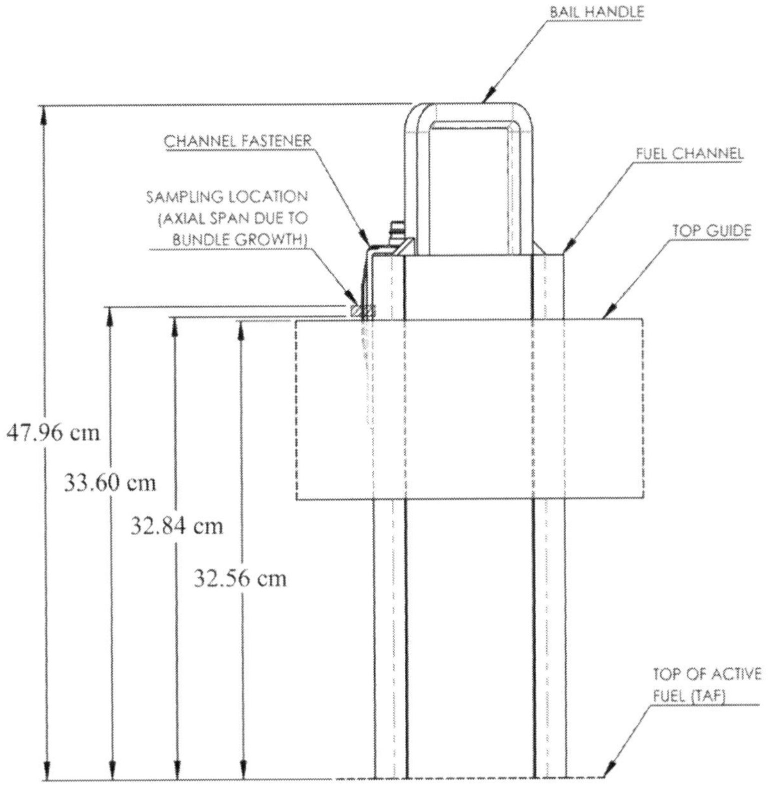

Fig. 6. Schematic representation of the channel fastener showing the locations where the fluence was calculated. The calculation accounted for channel length increase due to swelling.

As previously mentioned, two of the clips were irradiated in all three cycles (GEQ771/GEQ846), two were in C20 and C21 (GER212/GER238), and two were in C21 only (GES366/GES547). For BWRs, the average fuel cycle is 24 months. Other than bundle GER212, all of the bundle clips used for dosimetry were shuffled to a new location each cycle. As a result, the clips were exposed to widely varying water densities, thereby providing a very good CFD model benchmarking opportunity. Accounting for the Mn54 half-life of ~300 days, the C21 activation will be a good record of steam distribution effects on the flux history. However, more than 80% of the activity from the previous cycle will have decayed. The power history was weighted to account for the time varying effects on the dosimetry. The benchmarking is properly viewed as a one cycle benchmark with varying water density at each of the clips. In particular, during C21, two clips were located near core center, two were located at a position of 3 to 4 bundles in from the periphery, and two were located at the periphery.

Table 1. CFD model dosimetry benchmark results.

Dosimeter Reaction	Measured (M) Activity (dps/mg)	Calculated (C) Activity (dps/mg)	Ratio (C/M)
Fasteners Introduced in C21			
GES366S Fe54(n,p)Mn54	34,042.1	37,179.4	1.092
GES366I Fe54(n,p)Mn54	31,673.2	37,179.4	1.174
GES547S Fe54(n,p)Mn54	24,667.3	24,451.6	0.991
GES547I Fe54(n,p)Mn54	23,373.2	24,451.6	1.046
Fasteners Introduced in C20			
GER238S Fe54(n,p)Mn54	41,395.0	42,581.8	1.029
GER238I Fe54(n,p)Mn54	36,850.8	42,581.8	1.156
Fasteners Introduced in C19			
GEQ771S Fe54(n,p)Mn54	13,674.0	12,858.7	0.940
GEQ771I Fe54(n,p)Mn54	11,929.1	12,858.7	1.078
GEQ846S Fe54(n,p)Mn54	22,857.3	23,005.3	1.006
GEQ846I Fe54(n,p)Mn54	20,802.4	23,005.3	1.106
		Average Iron Result	1.062

4 Uncertainty analysis for above core structures

The fluence uncertainty at upper internals and vessel nozzles above TAF has contributions in addition to those typically considered in the core region. These added uncertainty contributors are:

- Additional transport methodology uncertainty in upper and lower regions
- Above core steam density uncertainty
- Additional steel transport uncertainty
- Additional core top and bottom edge power uncertainty
- Weld and nozzle location uncertainty

The above core steam density uncertainty estimates in Table 2 were derived from data generated from CFD parametric studies. The deterministic model improvements, such as buoyancy and compressibility, are not included in Table 2 because there are no uncertain inputs involved.

Combining all of the uncertainties in the active fuel region, as well as those above TAF, yields the following results:

- Horizontal/vertical shroud weld fluence uncertainties range from 13.3% to 19.0%
- Vessel plates, welds, and nozzles above TAF range from 17.4% to 19.4%.

5 Conclusions

The uncertainty analysis demonstrates that the MPM calculational methods provide fluence results within allowable tolerance bounds (±20%) for the reactor vessel, shroud welds, and surveillance capsules which lie within the axial active fuel region. This satisfies the requirements of RG 1.190. However, RG 1.190 does not address uncertainty and benchmarking outside this region. Nevertheless, analytic uncertainty evaluations of structures outside the beltline region have been made, and, especially above the core, uncertainties are within the RG 1.190 desired ± 20%. This low uncertainty level was only achievable by inclusion of the CFD model within the transport code suite. Retrospective dosimeters were cut from fuel channel clips and these data were used to benchmark the use of CFD data in the transport model. The average C/M ratio for the 10 dosimeters is 1.06. For each dosimetry result, it has been shown that all of the calculated dosimeter activities fall within ± 20% of the measurement. This meets the criterion set by RG 1.190 for acceptability of the calculations in the active fuel region.

Table 2. CFD water density overall uncertainty.

Uncertainty Contributor	Units	Best Case Input Used	Perturbed Input	Average Percent Difference (%)	σ Level	1σ Uncertainty (%)
Downcomer Water Temperature for Heat Transfer Through the Top Guide Wall	°C	275.1	287.8	5.73	3	1.91
Water/Steam Heat Transfer Coefficient	W/(m^2°C)	3123	4259	5.04	3	1.68
Water Droplet Size	Microns	1.0	2.3	6.36	3	2.12
Drag Coefficient	None	0.44	2.732	0.65	1	0.65
Mesh Refinement (Maximum Element Size in Mesh)	mm	60.76	30.87	1.90	1	1.90
Convergence	Iterations	100	400	1.80	3	0.60
TOTAL						3.92

References

1. US NRC Regulatory Guide 1.190, 'Calculational and Dosimetry Methods for Determining Pressure Vessel Neutron Fluence' , USNRC, Rockville, MD, (2001).
2. RSICC Computer Code Collection, CCC-543, TORTDORT-PC, Two- and Three-dimensional Discrete Ordinates Transport Version 2.7.3, Available from the Radiation Safety Information Computational Center, Oak Ridge National Laboratory, Oak Ridge, TN, (1996).
3. CASMO/SIMULATE, Studsvik's Advanced Three Dimensional Two-Group Reactor Analysis Code, Studsvik of America, Inc. (2020)

Validation of the Serpent 2 Monte Carlo code for reactor dosimetry applications

Eric Dorval[1,*]

[1]VTT Technical Research Centre of Finland, P.O. Box 1000, FI-02044, Finland

Abstract. The verification and validation (V&V) of the Serpent 2 Monte Carlo code for 3-D reactor dosimetry applications is being carried out at VTT. Two code-to-code computational benchmark cases were calculated by MCNP and Serpent 2. The computational efficiency of different variance reduction techniques was appraised. The validation part currently includes two experimental benchmarks from the SINBAD database: the Pool Critical Assembly-Pressure Vessel Facility (PCA); and the H.B. Robinson-2 (HBR-2) Pressure Vessel Dosimetry Benchmark. All simulations replicate the benchmark source specifications at pin level thanks to the development of a flexible, geometry-independent, fixed source definition for reactor dosimetry applications. Good agreement was found in all benchmark cases.

1 Introduction

The Serpent 2 Monte Carlo code [1] was originally targeted at lattice physics applications. Subsequent developments have extended its capabilities to full-core calculations, photon transport, activation and shielding. The V&V of Serpent applied to reactor dosimetry is being carried out mainly following the guidelines of the U.S. NRC Regulatory Guide 1.190 [2], which prescribes code-to-code calculational benchmarks as well as comparisons with benchmark measurements.

Two PWR verification cases were selected: a specification with Standard Core Loading Pattern (STD); and a Partial Length Shield Assembly (PLS) core [3]. Fast fluxes were calculated with MCNP [4] version 6.1 and Serpent 2, based on identical geometry and material definitions. The present validation comprises two experimental benchmarks: the PCA Pressure Vessel Facility; and the HBR-2 Vessel Dosimetry Benchmark. Detailed Serpent models were setup for the calculation of selected dosimetry reaction rates using cross section (XS) data from IRDFF [5] version 1.05.

* Corresponding author: eric.dorval@vtt.fi1)

2 Methods

2. 1 Fixed neutron source routines for reactor dosimetry applications

In-house modifications were made to the Serpent code (version 2.1.32) to allow fixed neutron source definitions of practical interest for reactor dosimetry. Source specifications are read from an external text file containing all the parameters needed by Serpent to sample source points that replicate the degree of detail/complexity imposed by the benchmarks. Sources from different core simulators for diverse reactor core types can be generated by using the appropriate format setup of this external file. Source points are sampled at two levels: assembly and pin. The fixed source specification is independent of the geometry definition. Table 1 summarizes the fixed neutron source features.

Table 1: Fixed neutron source options available in the new routines. Axial distributions may have different axial discretization per assembly. Energy distributions can be combined freely.

1st level (assembly)	2nd level (pin)	Pin type	Axial	Angular	Energy
Cartesian, hexagonal	Cartesian, hexagonal	square, (hollow) cylindrical, rectangular, (bent) MTR plate	piece-wise constant	isotropic	histogram, Watt fission spectrum, Maxwell fission spectrum

2.2 Benchmarks

Benchmark descriptions are left outside this article due to size considerations. The interested reader is referred to the appropriate documentation. Here, only the most salient features, assumptions and other indispensable information related to the solution method are given. Figure 1 (upper row) depicts top views of the reactor models studied in this work, whereas Figure 1 (bottom row) shows the corresponding importance maps generated with the built-in response matrix solver of Serpent [6] for the same cases.

Figure 1: Top views of the Serpent models used in this work. From left to right: STD; PLS; PCA (partial extents); and HBR-2. Bottom row: detector importance maps. Neutron importance increases from the blue tones toward the red ones. Figures are not to scale.

The importances calculated by Serpent's built-in deterministic solver are inversely proportional to the lower weight bounds used in weight windows. Importance maps should not be confused with *cell importances* (as in MCNP). Although Serpent can handle cell importances, importance maps are the way in which Serpent implements weight windows.

2.2.1 Computational benchmarks

MCNP input files of the STD and PLS numerical benchmark problems were facilitated by one of the authors of NUREG/CR-6115 [3]. Corresponding Serpent inputs were created from scratch. In addition to identical geometry, material, and fixed source definitions, the Serpent 2 models also feature the same Variance Reduction (VR) techniques used in Ref. [3], namely *cell importances* and energy cut-off. The adoption of *cell importances* increases the complexity of the geometry definition because it necessitates individual cells where the *cell importances* are set. This can be observed in the peripheral assemblies of the leftmost pictures in Figure 1 (left).

While the original models from NUREG/CR-6115 were based on multi-group cross section (XS) neutron transport libraries, this work was carried out with the continuous energy library ENDF/B-VII.1. Fast (E > 1 MeV) neutron fluxes computed by MCNP version 6.1 at two axial locations ("Peak" and "Weld") were taken as a reference solution. Given the vast number of fluxes to be compared, the agreement between Serpent and MCNP solutions is reported in the form of reduced chi-squared values:

$$\frac{\chi^2}{DOF} = \frac{1}{n}\sum_{i=1}^{n}\frac{(s_i - m_i)^2}{\sigma_i^2}, \tag{1}$$

where s_i and m_i denote the Serpent and MCNP results, respectively; n is the number of results, or degrees of freedom (DOF); and σ_i is an appropriate measure of combined uncertainty between MCNP and Serpent ($\sigma_{m,i}$ and $\sigma_{s,i}$, respectively):

$$\sigma_i = \sqrt{\sigma_{m,i}^2 + \sigma_{s,i}^2}\,. \tag{2}$$

Additional STD and PLS cases not present in Ref. [3] were calculated with Serpent. In these, the VR was switched from *cell importances* to one-group weight windows (WWs) generated by Serpent. Since the new source routines in Serpent do not require explicit definition of the geometry at pin level, yet two more additional cases were run, where every fuel assembly was represented as a single cell, whilst still retaining the full detail of the pin-wise source dependence. The aim of the latter cases is to quantify the computational overheads associated with the definition of unnecessary additional cells and the use of surface-tracking instead of delta-tracking [7]. The overheads are quantified through the figure of merit (FOM), defined for a given tally as:

$$FOM = \frac{1}{\sigma^2 T}, \tag{3}$$

where T is the total computation time, and σ is the mean absolute statistical uncertainty of all azimuthal detector locations considered: $\sigma = \widehat{\sigma_{s,i}}$.

2.2.2 PCA

The first experimental case of this work is the PCA Pressure Vessel Wall Facility Benchmark Configuration 12/13 [8,9], which features a fresh, highly enriched (93 %), materials testing reactor (MTR) core with 18 bent fuel plates per standard fuel element. The benchmark in SINBAD [10] includes experimental data for six nuclear reactions at seven ex-core locations in water, carbon steel, and air. Dosimeter locations denoted by "A1" ... "A7" are defined along the core horizontal centreline. The numbering increases with the distance to the PCA window. This window is situated to the left of the core (Figure 1, centre picture).

Two simplifications were introduced in the Serpent model: control rods (CRs) were not explicitly represented; and all fuel elements were modelled with 18 fuel plates (except for those with CR channels). The first assumption is well justified, since the fixed neutron source is a part of the benchmark specification, whereas the effect of the second simplification is expected to be small, because the fuel meat densities were adjusted to match the actual ^{235}U mass in elements with either 17 or 19 plates. This change was driven by the current implementation of the fixed source routine in Serpent, which is limited to the same number of pins (or plates) per assembly (element) at the 2nd level in Table 1.

The Serpent model was run using two neutron transport nuclear data libraries, namely ENDF/B-VII.1 [11] and JEFF-3.1.2 [12]. An unperturbed Watt fission spectrum of thermal neutrons on ^{235}U was adopted. Dosimetry XS data from IRDFF version 1.05 was used. One-group WWs were generated by Serpent. The results are presented in the form of calculated-to-experimental (C/E) reaction rate ratios per isotope and per location.

2.2.3 HBR-2

The HBR-2 benchmark [13] is the most complex one studied in this work. The complexity stems from the pin-wise nature of the neutron source produced by a core simulator, the time-dependence of the global (proportional to total power) and local (spatial) neutron source distributions, and the number of structural components. The SINBAD database [14] provides nine full-core, pin-wise neutron source specifications: one for cycle-average conditions; and others at eight selected time steps during cycle 9.

The Serpent model features explicit heterogeneous fuel elements with axially-uniform burnup distributions at mid-cycle conditions. Fuel inventories were calculated also with Serpent, based on a simplified, 2-D single-assembly model. This auxiliary depletion calculation supplied the data necessary to setup the neutron source rigorously, i.e., to allow sampling of isotope-dependent fission spectra weighted with the number of emitted neutrons, as opposed to fission fractions. Moreover, this source treatment provided estimates of the neutron source strength as a function of the cycle step. The neutron emission energies were sampled from Watt fission spectra corresponding to ^{235}U, ^{238}U and ^{239}Pu. The relative contributions varied according to the assembly exposure.

The VR techniques used were energy cut-off and one-group WWs. Dosimetry and transport XS data were taken from the IRDFF and ENDF/B-VII.1 library, respectively. The results of the benchmark are supplied in the form of C/E activity ratios per isotope as a function of the dosimetry location (surveillance capsule or cavity) for the three approaches adopted to cater for neutron sources during cycle 9, to be described in the Results section.

3 Results

3. 1 Computational benchmarks

Fast neutron fluxes calculated by MCNP and by Serpent are depicted in Figure 2. There is evidence of very good agreement when Serpent uses the same VR techniques as MCNP, namely *cell importances* and energy cut-off. This scenario is labelled as "**Case A**". Two additional scenarios were simulated with Serpent only: in **Case B**, *cell importances* were replaced by weight windows. In **Case C**, the geometry was simplified by merging all peripheral pin cells into single assembly cells; delta-tracking and weight windows were used. Table 2 summarizes the results for all the cases by reporting reduced chi-squared values through Equation (1) and FOMs via Equation (3) for the Serpent cases only.

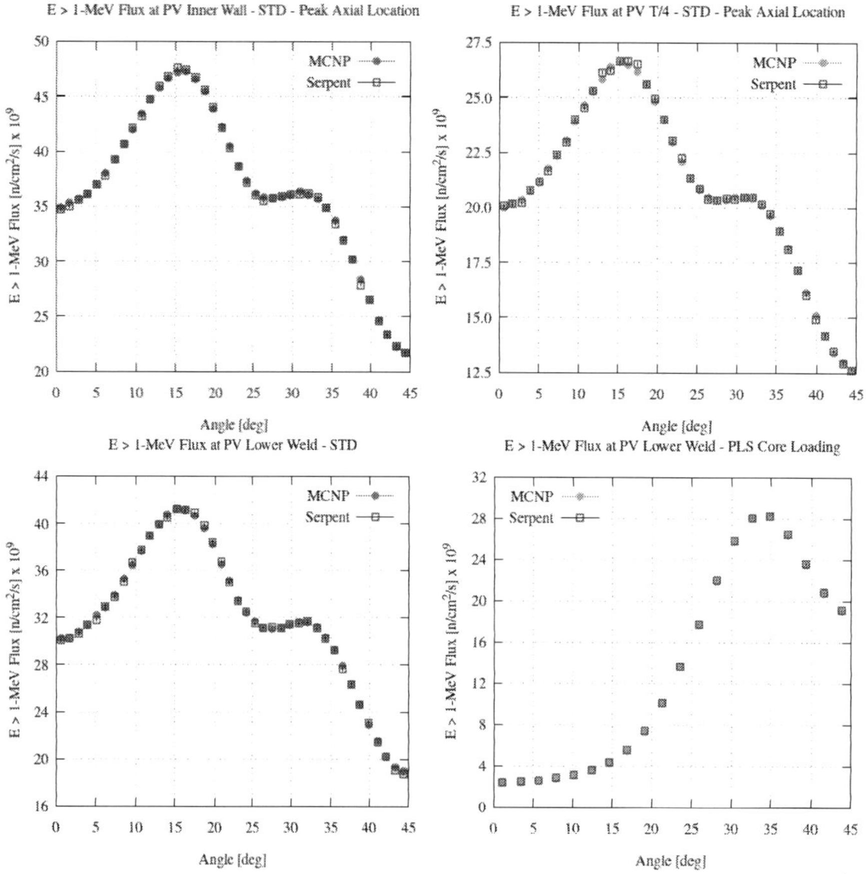

Figure 2: Fast neutron fluxes (and their statistical uncertainties) at various locations for the STD and PLS computational benchmarks using identical geometry and VR techniques (**Case A**).

In general, reduced χ^2 values in the interval [0.3, 2.0] are evidence of good statistical agreement. All combinations of geometry definitions, tracking and VR techniques yield results comparable to those of MCNP. The FOM values of **Cases B** and **C** underline the higher computational efficiency of WWs against *cell importances* for this case, as well as the improvements brought about by the geometry simplifications and delta-tracking.

Table 2: Reduced χ^2 and FOM for computational benchmarks. Azimuthal fast neutron flux profiles were calculated at the Pressure Vessel (PV) Lower Weld, Inner Wall, and fourth thickness (T/4).

Result	Case A		Case B		Case C	
	χ^2/DOF	FOM	χ^2/DOF	FOM	χ^2/DOF	FOM
STD PV Lower Weld	0.9	11	1.2	65	0.6	123
STD PV Inner Wall – Peak	1.2	47	1.0	248	1.0	452
STD T/4 – Peak	0.8	15	1.6	79	0.9	148
PLS Lower Weld	1.7	316	0.3	449	0.8	1542

3.2 Experimental benchmarks

3.2.1 Pool Critical Assembly – Pressure Vessel Facility Benchmark

Figure 3, Table 3 and Table 4 present the C/E reaction rate ratios along with their statistical uncertainties, which are largely governed by the experimental data. The overall agreement of Serpent results with experiments is very satisfactory. No significant differences based on the adoption of either ENDF/B-VII.1 or JEFF-3.1.2 data are evident. The high uncertainty from the ^{237}Np dosimeter at location A7 stems from the measurements (8.4 %).

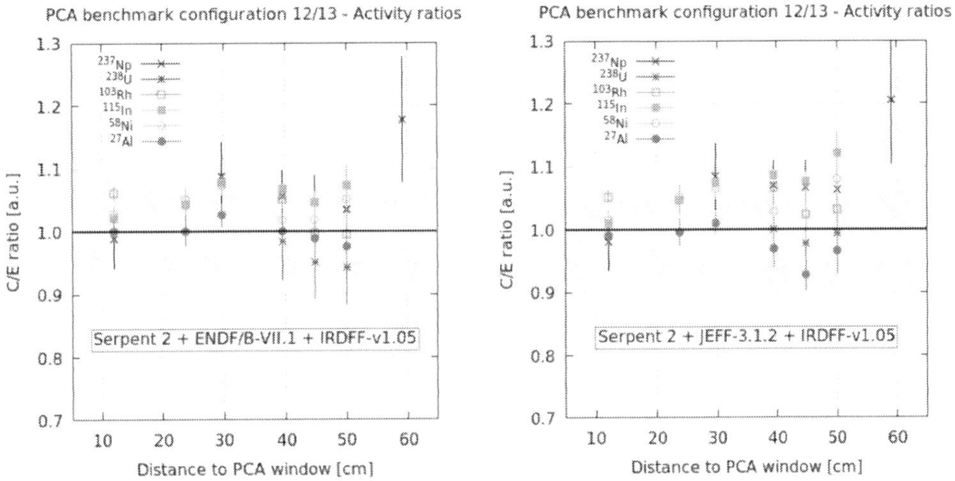

Figure 3: Specific activity C/E ratios for the PCA case. Calculations with Serpent 2 were conducted using transport XS data from ENDF/B-VII.1 (left) and JEFF-3.1.2 libraries (right). Both cases use the same dosimetry XS data (IRDFF version 1.05). Shaded areas represent a ± 10 % uncertainty range.

Table 3: C/E specific activity ratios for the PCA benchmark problem. Uncertainties of averages are based on observed sample standard deviations only. Transport XS data from ENDF/B-VII.1.

Location / Distance (cm)	ENDF/B-VII.1 XS data											
	^{237}Np(n,f)^{137}Cs		^{238}U(n,f)^{137}Cs		^{103}Rh(n,n')^{103m}Rh		^{115}In(n,n')^{115m}In		^{58}Ni(n,p)^{58}Co		^{27}Al(n,α)^{24}Na	
	C/E	unc	C/E	unc	C/E	unc	C/E	unc	C/E	unc	C/E	unc
A1 / 12.0	0.99	0.05	-	-	1.06	0.01	1.02	0.01	1.03	0.01	1.00	0.01
A2 / 23.8	-	-	-	-	-	-	1.04	0.02	1.05	0.02	1.00	0.02
A3 / 29.7	1.09	0.05	-	-	-	-	1.08	0.01	1.07	0.02	1.03	0.02
A4 / 39.5	1.06	0.04	0.98	0.06	1.05	0.02	1.07	0.01	1.02	0.01	1.00	0.03
A5 / 44.7	1.05	0.04	0.95	0.06	1.00	0.05	1.05	0.02	1.02	0.02	0.99	0.03
A6 / 50.1	1.03	0.05	0.94	0.06	1.00	0.05	1.07	0.03	1.05	0.02	0.98	0.03
A7 / 59.1	1.18	0.10	-	-	-	-	-	-	-	-	-	-
Average	**1.07**	**0.06**	**0.96**	**0.02**	**1.03**	**0.03**	**1.06**	**0.02**	**1.04**	**0.02**	**1.00**	**0.02**

Table 4: C/E specific activity ratios for the PCA benchmark problem. Uncertainties of averages are based on observed sample standard deviations only. Transport XS data from JEFF-3.1.2.

Location / Distance (cm)	JEFF-3.1.2 XS data											
	^{237}Np(n,f)^{137}Cs		^{238}U(n,f)^{137}Cs		^{103}Rh(n,n')^{103m}Rh		^{115}In(n,n')^{115m}In		^{58}Ni(n,p)^{58}Co		^{27}Al(n,α)^{24}Na	
	C/E	**unc**	**C/E**	**unc**	**C/E**	**unc**	**C/E**	**unc**	**C/E**	**unc**	**C/E**	**unc**
A1 / 12.0	0.98	0.05	-	-	1.05	0.01	1.01	0.01	1.01	0.01	0.99	0.01
A2 / 23.8	-	-	-	-	-	-	1.05	0.02	1.05	0.02	0.99	0.02
A3 / 29.7	1.08	0.05	-	-	-	-	1.07	0.01	1.07	0.02	1.01	0.02
A4 / 39.5	1.07	0.04	1.00	0.06	1.07	0.02	1.09	0.01	1.03	0.01	0.97	0.03
A5 / 44.7	1.07	0.05	0.98	0.06	1.02	0.05	1.08	0.02	1.03	0.02	0.93	0.03
A6 / 50.1	1.06	0.05	0.99	0.06	1.03	0.05	1.12	0.03	1.08	0.03	0.96	0.03
A7 / 59.1	1.21	0.10	-	-	-	-	-	-	-	-	-	-
Average	**1.08**	**0.07**	**0.99**	**0.01**	**1.04**	**0.02**	**1.07**	**0.04**	**1.04**	**0.03**	**0.98**	**0.03**

3.2.2 H.B. Robinson-2 Pressure Vessel Dosimetry Benchmark

Table 5 summarizes the C/E activity ratios at the capsule and cavity locations, respectively. These locations are depicted by green dots in Figure 1 (right). The three cases reported per location differ in how the time-dependence of the neutron source was accounted for. All three cases account for variations in reactor power when calculating activities. In **Case 1** and **Case 2**, a single transport calculation was executed. In **Case 2**, reaction rates were corrected based on the power of the peripheral fuel assemblies at eight time-steps, as done somewhere else [13,14]. **Case 3** is the result of post-processing eight different Serpent runs, each one featuring a distinctive neutron source as supplied in the benchmark specification: atomic reaction rates from Serpent runs were associated with their corresponding irradiation history intervals. Then, decay-corrected activities in units of Bq/mg were obtained.

Table 5: C/E ratios of activities at End-of-Irradiation (EoI) in HBR-2 benchmark. Averages report sample standard deviations only. Uncertainties below 0.5 % read "0.00".

Reaction	Capsule						Cavity					
	Case 1		Case 2		Case 3		Case 1		Case 2		Case 3	
	C/E	unc	C/E	unc	C/E	unc	C/E	unc	C/E	unc	C/E	unc
^{237}Np(n,f)^{137}Cs	1.00	0.00	0.98	0.00	0.97	0.00	0.85	0.00	0.81	0.00	0.81	0.00
^{238}U(n,f)^{137}Cs	0.97	0.00	0.95	0.00	0.94	0.00	0.94	0.00	0.90	0.00	0.91	0.00
^{58}Ni(n,p)$^{58m+g}$Co	0.96	0.00	1.01	0.00	0.97	0.00	0.85	0.01	0.94	0.01	0.93	0.02
^{54}Fe(n,p)^{54}Mn	1.01	0.00	1.02	0.00	1.00	0.00	0.97	0.01	0.97	0.01	0.98	0.01
^{46}Ti(n,p)^{46}Sc	1.02	0.00	1.06	0.00	1.03	0.00	0.95	0.01	1.04	0.01	1.02	0.02
^{63}Cu(n,α)^{60}Co	1.05	0.01	1.04	0.01	1.03	0.00	1.02	0.02	0.98	0.02	0.98	0.01
Average	**1.00**	0.00	**1.01**	0.00	**0.99**	0.00	**0.93**	0.01	**0.94**	0.01	**0.94**	0.01
(without ^{237}Np)	**1.00**	0.00	**1.02**	0.00	**0.99**	0.00	**0.95**	0.01	**0.96**	0.01	**0.96**	0.01

In all cases, the agreement with experiments is satisfactory, and there is no evidence of drastic improvements when opting for either **Case 2** or **Case 3**, except for the reaction ^{58}Ni(n,p)$^{58m+g}$Co. It is worthwhile mentioning that the results reported in this work only include the (propagated) statistical uncertainties in Serpent tallies, since the benchmark does not report experimental uncertainties. Corrections due to photo-fission and activation of impurities were applied to the experimental data prior to its comparison against calculations. Average results are presented with and without the contribution of the $^{237}Np(n,f)^{137}Cs$ reaction, since there likely is an error in the measured ^{137}Cs activities [13].

4 Conclusions

A systematic V&V of the Serpent 2 Monte Carlo applied to reactor dosimetry is being carried out according to U.S. NRC guidelines. Thus far, two computational and two experimental benchmark models were analysed, yielding good agreement with reference data. The models include a high degree of detail, and it is envisaged that they could be used as a basis for future studies. In particular, special effort was invested in the development of fixed source routines in Serpent that could match as closely as possible the dosimetry benchmark specifications. Some of the cases studied allowed quantifying the effect of variance reduction techniques and modelling practices on computational efficiency. The insights gained during this work will be applied in the upcoming benchmark calculations to continue the verification and validation of Serpent 2.

Gabriel Zamonksy is dearly acknowledged for supplying the MCNP inputs (including neutron source specifications) for the PLS and STD cases. This work was funded by the Finnish Research Programme on Nuclear Power Plant Safety SAFIR2022/RACSA project.

References

1. J. Leppänen, M. Pusa, T. Viitanen, V. Valtavirta, and T. Kaltiaisenaho, Ann. Nucl. Energy **82**, pp. 142-150 (2015)
2. U.S. Nuclear Regulatory Commission, *Regulatory Guide 1.190* (2001)
3. J. Carew, K. Hu, A. Aronson, A. Prince, and G. Zamonsky, NUREG/CR-6115 (2001)
4. D. B. Pelowitz (Ed.), *MCNP6™ User's Manual*, LA-CP-13-00634 Rev. 0 (2013)
5. E. Zsolnay, R. Capote Noy, H. Nolthenius, and A. Trkov, INDC(NDS)-0616 (2012)
6. J. Leppänen, Nucl. Technol **205** (11), pp. 1416-1432 (2019).
7. J. Leppänen, Ann. Nucl. Energy **37**, pp. 715-722 (2010)
8. I. Remec and F. Kam, NUREG/CR-6454 (1997)
9. W. McElroy (Ed.), NUREG/CR-1861 (1981)
10. I. Kodeli, E. Sartori, and B. Kirk, Proc. Am. Nuc. Soc. 14th Topical Meeting of the Radiation Protection and Shielding Division (2006)
11. National Nuclear Data Center (NNDC), ENDF/B-VII.1 Evaluated Nuclear Data Library, Brookhaven National Laboratory (2011)
12. Nuclear Energy Agency (NEA), Joint Evaluated Fission and Fusion nuclear data library JEFF-3.1.2 (2016)
13. I. Remec and F. Kam, NUREG/CR-6453 (1997)
14. R. Orsi, Nucl. Eng. Technol.**52** (2), pp. 448-455(2020)

Method for calculating delayed gamma-ray response in the ACRR Central Cavity using MCNP

Melissa Moreno[1,*], and *Edward* Parma[2]

[1]Sandia National Laboratories, Albuquerque, NM, USA
[2] Sandia National Laboratories (Retired), Albuquerque, NM, USA

Abstract. This document presents the process for a new method developed for the characterization of the delayed gamma-ray radiation fields in pulse reactors like the Annular Core Research Reactor (ACRR). The environments used to test this method in the ACRR Central Cavity were various standard irradiation configurations that used various amounts of moderators and shielding. All environment configurations used the same fission product gamma-ray source energy spectrum. This method required the use of fission sites recorded in the Monte Carlo N-Particle (MCNP) KCODE source tape. A FORTRAN script was written to translate and extract the coordinates for the fission sites. The fission sites were then input into an MCNP source mode input file. Using a MATLAB script, a parametric analysis was performed and determined that 10K fission sites are an appropriate number of coordinates to converge to the correct answer. The method gave excellent results as compared to previous methods. This method can be applied to other pulse research reactors as well.

1 Introduction

To maintain a high degree of fidelity when performing qualification tests at the Annular Core Research Reactor (ACRR), the irradiation environments in the cavity experiment region must be characterized. Environment characterization includes determining the neutron, prompt gamma-ray, and delayed gamma-ray energy spectra and intensity. To characterize the neutron and gamma-ray environments, both experimental and computational modeling work must be performed. To determine the neutron energy spectrum, activation foils are irradiated at the experiment location. The foil activity results are compared to the results attained from the computed neutron energy spectrum. Prompt and delayed gamma-ray energy spectra

* Corresponding author: mmoren@sandia.gov
Sandia National Laboratories is a multimission laboratory managed and operated by National Technology & Engineering Solutions of Sandia, LLC, a wholly owned subsidiary of Honeywell International Inc., for the U.S. Department of Energy's National Nuclear Security Administration under contract DE-NA0003525. This paper describes objective technical results and analysis. Any subjective views or opinions that might be expressed in the paper do not necessarily represent the views of the U.S. Department of Energy or the United States Government.

cannot be measured experimentally, but the intensities can be determined using passive and active dosimetry measurement techniques. For photon energies above the K-edges for photon cross sections (~80 keV), most material response functions fail to show any strong energy-dependent structure that can be used to resolve the spectrum. There is simply no effective way to discriminate energy groups in a gamma-ray fluence within a small volume in a reactor core. Integral dose measurements can be performed and are used as benchmarks for calculated integral responses. These integral measurements have value in validation but are not useful for spectrum adjustment. However, both passive dosimeters (e.g., TLDs) and active dosimeters (e.g., calorimeters) have some neutron and delayed gamma-ray contribution effects that make interpreting the results much more complex [1].

Prompt gamma-rays are identified as those produced immediately from fission, radiative capture, or inelastic scattering. These gamma-rays are considered to be instantaneously generated along with the neutrons. Using the particle transport Monte Carlo N-Particle (MCNP) [2], the prompt gamma-ray transport can be calculated in conjunction with the neutrons using KCODE mode. This allows for a high-fidelity calculation to be performed and the prompt gamma-ray energy spectrum and intensity to be calculated at the irradiation location of interest simultaneously with the neutrons.

Delayed gamma-rays are identified as those produced from the decay of the fission products and by the activation of materials in the core. For this report, and the previous characterization reports, the fission produced beta particles and their associated bremsstrahlung were not analyzed. The activation delayed gamma-rays were also not considered in this analysis; only fission delayed gamma-rays were considered in this study. Future work may look at the fission product beta particles and activation gamma-rays. For delayed gamma-rays, there is not a simple method to calculate the transport using MCNP. In theory, the fission locations that are generated in a KCODE problem could be saved and used in a source problem using a delayed gamma energy spectrum. However, MCNP does not allow for a methodology to do this directly. Previous methods, used in the characterization reports, used a simplistic row-by-row fission tally in a KCODE mode as a source distribution. The row-by-row fission tally results were used in a source problem with a defined gamma-ray energy spectrum. This methodology was known to be deficient because it did not account for any axial or internal fuel geometric dependency.

The method proposed in this work was to determine the actual fission sites from the KCODE problem, and then use those fission locations in a source problem with a defined gamma emission energy spectrum.

2 Methodology

The neutronics ACRR model used to test this method was developed for MCNP [3]. This paper expands only on the Free Field (FF) environment for the ACRR, but there were eight total environments used for the testing of this method [4]. For the ACRR, the environments tested were FF [1], 44-Inch Lead-Boron (LB44) [5], Polyethylene-Lead-Graphite (PLG) [6] and Cadmium-Polyethylene (CdPoly) [7], and for the FREC-II, the environments tested were FF with rods-up and rods-down [8] and CdPoly with rods-up and rods-down. The unique MCNP models for the analyses are found in the appendices of the characterization reports [1,3-7].

The basic concept for calculating the delayed gamma-ray response for this work was to determine the actual fission sites from an MCNP KCODE calculation, and then use those locations to run a separate source problem with a defined gamma emission energy spectrum.

First, an MCNP KCODE mode calculation for each of the eight environment configurations was done. This mode creates a binary source tape or SRCTP file which at the end of the MCNP run contains the most updated fission sites. The SRCTP files were translated to readable data and all the x, y, z coordinates were stored in a separate file. From these files, 10,000 fission sites were selected and inserted into another MCNP file but this time using the SDEF mode. This mode is the general source definition in MCNP. For this experiment the source was defined by a probability distribution requiring a SI (Source Information) and a SP (Source Probability) card. The selected 10,000 fission sites all had equal probability of being selected to be the starting point for the delayed gamma calculations.

The fission-product delayed gamma-ray energy spectrum used was for fast U-235 fission in the time interval from 0.2 to 0.5 seconds as given in Engle and Fisher [9] [10]. This spectrum was used since the ACRR is an epithermal system and approximately 87% of the neutron population is above 1 eV and 65% above 1 keV [1]. The total delayed gamma-ray energy released from the fission products used was 6.33 MeV per fission for U-235 fission from ENDF/B-VII.1 [11]. This value is about half of the value emitted as prompt gamma-rays, and therefore cannot be ignored as a trivial quantity. The number of delayed gamma-rays emitted per fission was then calculated to be 6.57 delayed photons/fission, using the source delayed gamma-ray energy spectrum for fast U-235 fission.

Figure 1 shows the time-integrated energy fraction from the emission of the delayed gamma-rays on a per fission basis. Although fission products continue to decay over many years, most of the energy (~85%) is released from the shorter-lived products over 1E4 seconds (2.8 hours). The fractional energy released up to 0.1 s is ~1%, ~6.5% up to 1 s, ~23.2% up to 10 s, and ~55% up to 300 s (5 minutes) [12].

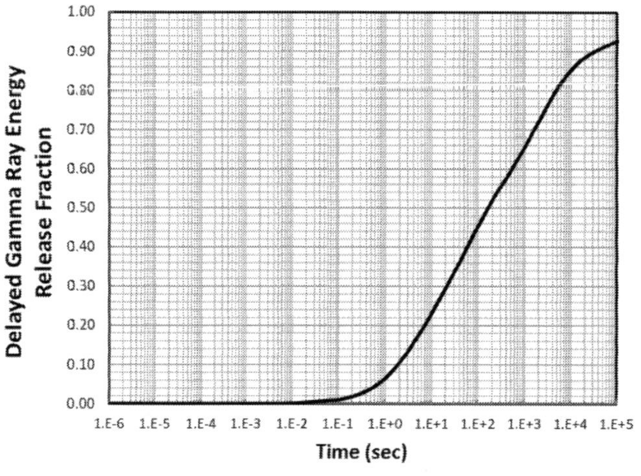

Figure 1. Time Dependent Delayed Gamma-Ray Energy Release Fraction from Fission.

3 Point Determination

When the source tape is generated in the KCODE MCNP calculation, it contains the coordinates for the fission sites up to the number of particles per cycle. The determination of how many fission sites to include in the SDEF code was an important question that required some parametric analysis. This analysis was approached by using the FF configurations for the ACRR. Random coordinates were selected in groups of 10, 100, 1K, 10K, 50K and 100K, and these were plotted in MATLAB. These plots were done in XY, XZ, and YZ coordinates.

Figure 2 shows the XY point results for the coordinates in the ACRR FF. The groups of 10 and 100 points were too sparse, making it nearly impossible to determine the shape of the reactor. The group of 1K points had a clearer shape but not quite as defined. Lastly, the 10K, 50K, and 100K point plots almost looked like they were solid drawn shapes. A similar distribution was found for FREC-II rods-up and rods-down [4].

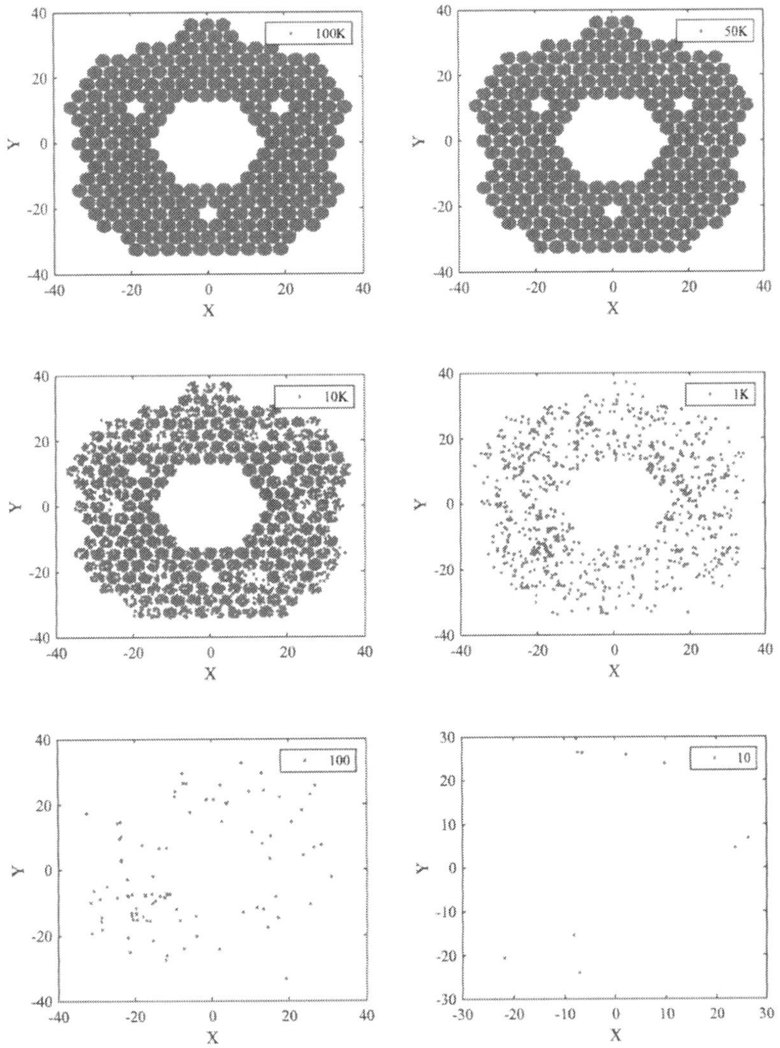

Figure 2. ACRR XY view of the coordinate distribution.

After running the MCNP SDEF for all the sets of points, the 48-energy group delayed gamma-ray fluence per MJ was plotted and is shown in Figure 3. These spectra were calculated using 5 billion particles resulting in an average total uncertainty of 0.05%. In this figure it is seen that the 1K, 10K, 50K, and the 100K results are all converging to the same curve and are higher in magnitude than the plot generated for the ACRR FF Characterization Report [1]. At the 0.5 MeV peak, the 10 Point was 112% below the characterized value, the 100 Points was 26% above the characterized value, 1K was 15% above the characterized value, 10K points was 14% above the characterized value, 50K points was 11% above the characterized value, and 100K points was 13% above the characterized value.

Since either 10K, 50K, or 100K seemed acceptable for this method, 10K points was selected for the ease of managing data. It was concluded that the same number of points would work for both the ACRR and the FREC-II since they had similar results for all the groups. Plots generated for the FREC-II can be seen in the full report [4].

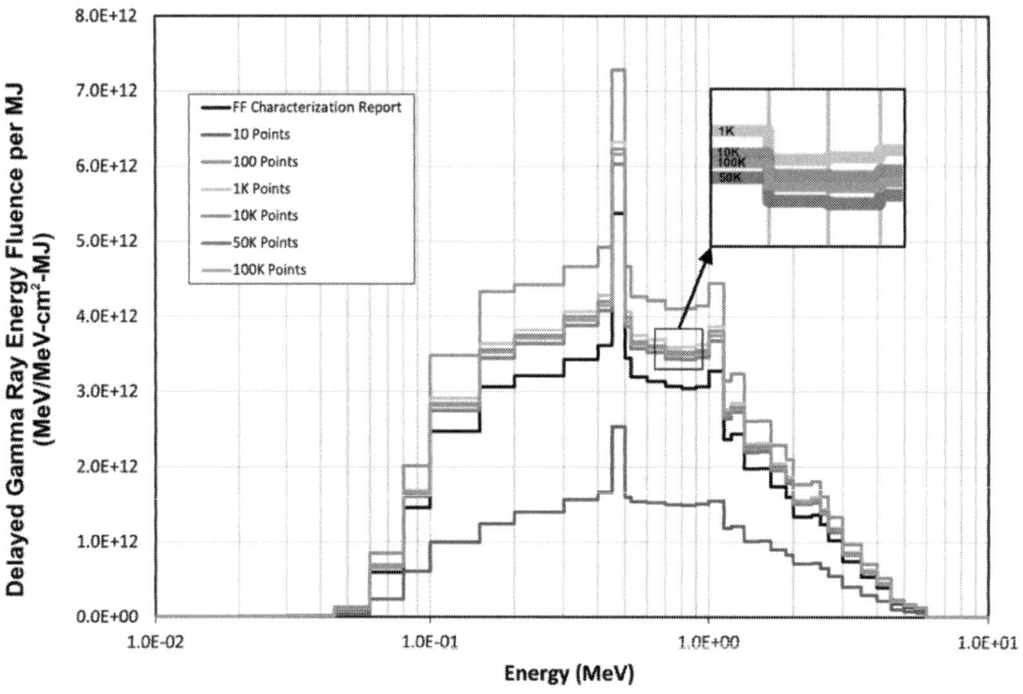

Figure 3. ACRR FF MCNP plots versus ACRR FF Characterization Report plot.

4 Delayed Gamma-Ray Spectrum Characterization

The MCNP 48-energy group delayed gamma-ray fluence is shown in Figure 4 for the FF ACRR using the 10K points. The delayed gamma-ray energy fluence has a prominent peak at ~0.5 MeV. This peak represents the electron-positron annihilation photon energy (0.511 MeV). The ACRR FF 10K plot shows a 14% higher intensity compared to the previous characterization report. This means the previous method was underestimating the delayed gammas generated due to fission in the ACRR, thus misrepresenting the delayed gamma spectra.

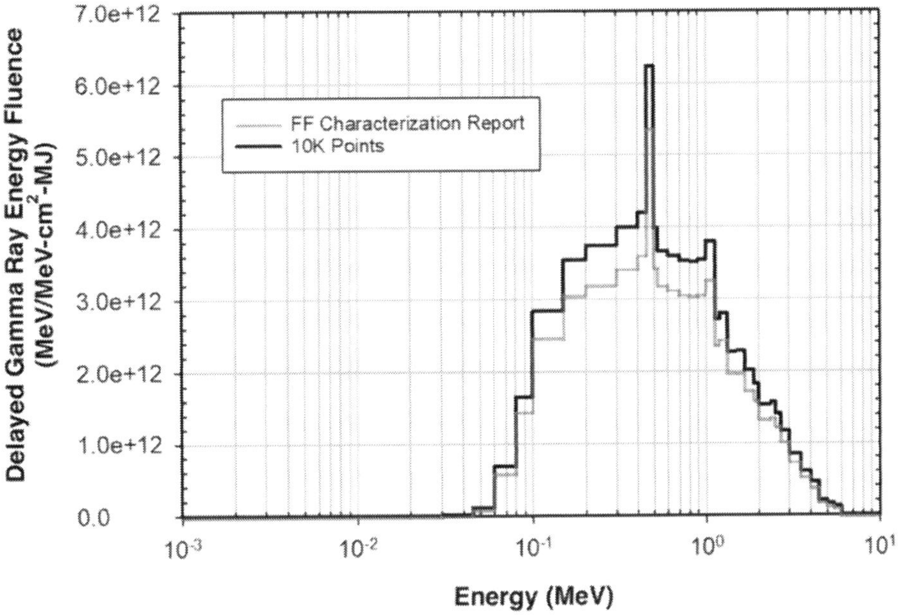

Figure 4. ACRR Free Field 10K plot versus previous report plot.

The MCNP ACRR models are run to calculate different tallies for each individual environment using the 48-energy group gamma-ray spectrum. These results and several integral metrics and conversion factors may be useful to the experimenter. Among the metrics calculated are Average delayed gamma-ray (dγ) energy (MeV), Fluence Conversion ([dγ/cm^2]/MJ), Total (Ionizing) Si Dose (rad[Si]/[dγ/cm^2]), Total (Ionizing) Carbon Dose (rad[C]/[dγ/cm^2]), and Total (Ionizing) CaF_2:Mn (TLD) Dose (rad[CaF_2:Mn]/[dγ/cm^2]). For a complete set of tables for all environments and conversion methods, refer to the full Sandia report [4].

Table 1 ACRR Metrics.

Metric	10	100	1K	10K	50K	100K
Average delayed gamma-ray (dγ) energy (MeV)	0.937	0.849	0.862	0.863	0.865	0.863
Fluence Conversion ([dγ/cm^2]/MJ)	4.30E+12	1.27E+13	1.09E+13	1.07E+13	1.04E+13	1.06E+13
Total (Ionizing) Si Dose (rad[Si]/[dγ/cm^2])	3.80E-10	3.49E-10	3.54E-10	3.54E-10	3.54E-10	3.57E-10
Total (Ionizing) Carbon Dose (rad[C]/[dγ/cm^2])	3.69E-10	3.38E-10	3.43E-10	3.43E-10	3.44E-10	3.47E-10
Total (Ionizing) CaF_2:Mn (TLD) Dose (rad[CaF_2:Mn]/[dγ/cm^2])	3.77E-10	3.47E-10	3.51E-10	3.52E-10	3.52E-10	3.55E-10

5 Conclusion

The method for calculating delayed gamma responses in characterized environments using this technique was successful. After calculating the decay gamma-ray fluence using MCNP for eight different environment configurations, the results showed that this procedure minimizes the assumptions made for the distribution function that was previously done. This method accounts for axial and internal fuel geometric dependencies by improving the sources being used, resulting in more accurate spectral information. This new method could become the formal approach to calculating delayed gamma-ray responses more accurately for any pulse reactor. Future work may look at the fission product beta particles and activation gamma-rays, and MCNP card investigations to assess if a more established method can be utilized to accomplish the same results without assumptions being made on the fission sources.

References

[1] E. J. Parma, G. E. Naranjo, R. M. Vega, L. L. Lippert, D. W. Vehar and P. J. Griffin, "Radiation Characterization Summary: ACRR Central Cavity Free-Field Environment with the 32-Inch Pedestal at the Core Centerline (ACRR-FF-CC-32-cl)," Technical Report SAND2015-6483, 2015.

[2] Los Alamos National Laboratory, "MCNP User's Manual Code Version 6.2," LA-UR-17-29981, 2017.

[3] K. R. DePriest, P. J. Cooper and E. J. Parma, "MCNP/MCNPX Model of the Annular Core Research Reactor," Technical Report SAND2006-3067, 2006.

[4] M. A. Moreno and E. J. Parma, "Method for Calculating Delayed Gamma-Ray Response in the ACRR Central Cavity and FREC-II Cavity using MCNP," Technical Report SAND2019-8746, 2019.

[5] E. J. Parma, T. J. Quirk, L. L. Lippert, P. J. Griffin, G. E. Naranjo and S. M. Luker, "Radiation Characterization Summary: ACRR 44-Inch Lead-Boron Bucket Located in the Central Cavity on the 32-Inch Pedestal at the Core Centerline (ACRR-LB44-CC-32-cl)," Technical Report SAND2013-3406 , 2013.

[6] E. J. Parma, D. W. Vehar, L. L. Lippert, P. J. Griffin, G. E. Naranjo and S. M. Luker, "Radiation Characterization Summary: ACRR Polyethylene-Lead-Graphite (PLG) Bucket Located in the Central Cavity on the 32-Inch Pedestal at the Core Centerline (ACRR-PLG-CC-32-cl)," Technical Report SAND2015-4844 , 2015.

[7] E. J. Parma, G. E. Naranjo, K. I. Kaiser, J. F. Arnold, L. L. Lippert, R. D. Clovis, L. E. Martin, T. J. Quirk and D. W. Vehar, "Radiation Characterization Summary: ACRR Cadmium-Polyethylene (CdPoly) Bucket Located in the Central Cavity on the 32-Inch Pedestal at the Core Centerline (ACRR-CdPoly-CC-32-cl)," Technical Report SAND2016-10114 , 2016.

[8] E. J. Parma, G. E. Naranjo, L. L. Lippert, R. D. Clovis, L. E. Martin, K. I. Kaiser, J. Emmer, J. Greenburg, J. O. Klein, T. J. Quirk, D. W. Vehar and P. J. Griffin, "Radiation Characterization Summary: ACRR-FRECII Cavity Free-Field Environment at the Core Centerline (ACRR-FRECII-FF-cl)," Technical Report SAND2017-8674, 2017.

[9] L. Engle and P. Fisher, "Delayed Gammas from Fast-Neutron Fission of Th232, U233, U235, U238 and Pu239," *Physical Review,* vol. 134, no. 4B, pp. B796-B816, 25 May 1964.

[10] L. Engle and P. Fisher, "Los Alamos Report LAMS-2642," 1962.

[11] M. B. Chadwick and et al., "ENDF/B-VII.1 Nuclear Data for Science and Technology: Cross Sections, Covariances, Fission Product Yields and Decay Data," *Nuclear Data Sheets,* no. 112, p. 2887–2996, 2011.

[12] T. K. Lane and E. J. Parma, "Delayed Fission Gamma-ray Characteristics of 232Th, 233U, 235U, 238U, and 239Pu," Technical Report SAND2015-7024, Albuquerque, 2015.

Validation of Neutron Flux and Activation in the Nagra AMAC model of the Mühleberg Power Plant

Valentyn Bykov[1,*], *Ben* Volmert[1], *Sylvain* Pelloux[2], and *Erwin* Neukäter[2]

[1] Nagra (National Cooperative for the Disposal of Radioactive Waste), Hardstrasse 73, 5430 Wettingen, Switzerland
[2] BKW Energie, Kernkraftwerk Mühleberg, 3203 Mühleberg, Switzerland

Abstract. The decommissioning of the Mühleberg Nuclear Power Plant (KKM) is supported by Nagra (National Cooperative for the Disposal of Radioactive Waste) through computational analysis of neutron activation in the reactor components and structures using the Nagra Advanced Methodology for Activation Characterization (AMAC). To validate these calculations, measurements of neutron flux and neutron-induced activity were carried out in the form of two foil activation campaigns (72 samples) and three concrete sampling campaigns (118 samples). Generally, the ratio of calculated to measured activities is in the order of 2 in the proximity of the reactor pressure vessel and 3 in the upper sections of the drywell. As expected, the flux in the lower (spherical) part of the drywell is significantly overestimated due to the simplified modelling of this space. The overestimation, which ratios of up to 10, can be corrected by scaling the calculated radionuclide activity distributions in the vicinity of the validation locations.

1 Introduction

The Mühleberg Power Plant (KKM) is a General Electric Type-4 boiling water reactor (BWR/4) with Mark I containment, located in Switzerland. KKM was shutdown on December 20, 2019 and is presently being decommissioned. Nagra, the Swiss national cooperative for the disposal of radioactive waste, is supporting this project by applying its Advanced Methodology for Activation Characterization (AMAC) to characterize the neutron activation within the reactor internals, the reactor pressure vessel (RPV), and the surrounding structures. This work represents the first look at the comprehensive results of the measurement campaigns carried out to validate the neutron flux and activation calculated with AMAC.

* Corresponding author: valentyn.bykov@nagra.ch

2 Nagra Advanced Methodology for Activation Characterization (AMAC)

In order to quantify the amount of radioactive waste arising from nuclear power plant decommissioning, Nagra utilizes AMAC [1], based on three-dimensional MCNP [2] models of the plant.

Inside the RPV, the geometry is well known. The structures are typically well documented and only rarely change during the lifetime of the plant. Conversely, outside of the RPV, the information is much scarcer. There are generally no records of the objects filling the surrounding room over the lifetime of a plant, let alone their exact dimensions, weight, and material compositions. Many of the objects also tend to be moved, added, or removed. This poses a challenge for the accurate modeling of neutron transport in the plant.

Filling the room with a homogenized mixture of a typically expected amount of material (representing the objects inside the room) has been observed to lead to an incorrect neutron spectrum and overestimation of shielding provided by the objects. This is due to the fact that in reality, neutrons can pass in between the various objects (which occupy only a fraction of the space inside the room). Conversely, when a homogenized mixture fills the room, all neutrons are guaranteed to encounter (albeit diluted) shielding material. For this reason, a conservative assumption is made for all MCNP models used for AMAC: rooms are assumed to be empty. This unavoidably leads to an overestimation of the neutron flux (due to the absence of any shielding materials along the neutron path in these rooms). However, this conservative assumption is deemed preferential to an underestimation of the neutron flux. Figure 1 shows the MCNP model of KKM; it can be seen that the drywell is modeled as empty.

The effect of this overestimation typically grows with distance from the neutron source, as more and more shielding material is neglected along the path in those empty rooms. Therefore, a comparison of calculated quantities against measurements is highly desirable, as it allows the determination of the degree of flux overestimation in the area. Since the AMAC modeling of the power plants also involves other assumptions and simplifications, it is also desirable to validate the quantities calculated for the structures close to the core (where the empty room assumption isn't the leading source of error).

Validation of AMAC calculations has typically been achieved in two ways. The first approach, typically done when the power plant is still operating, validates the neutron flux using in-situ foil activation. The second approach, mostly done following the final shutdown of the plant, validates the activity of the neutron activation induced radionuclides within plant components and structures against measurements.

3 Flux Validation with Activation Foils

Neutron transport is validated by in-situ sample activation campaigns, in which small metal foils, stacked inside cylindrical containers, are placed at key locations during a refueling outage and removed during the subsequent outage, thereby being activated during one whole reactor cycle. The activity of activation products is measured with gamma spectrometry and compared with the values calculated using the MCNP model.

Two separate foil-activation campaigns have been carried out at KKM. The first campaign took place during reactor cycle 33 (2005-2006) and involved 50 samples, including two chains of 18 and 14 samples, respectively, installed on the inner and outer surface of the bioshield. The second campaign took place during reactor cycle 39 (2012-2013) and involved 22 samples. The locations of all these samples are summarized in Figure 1.

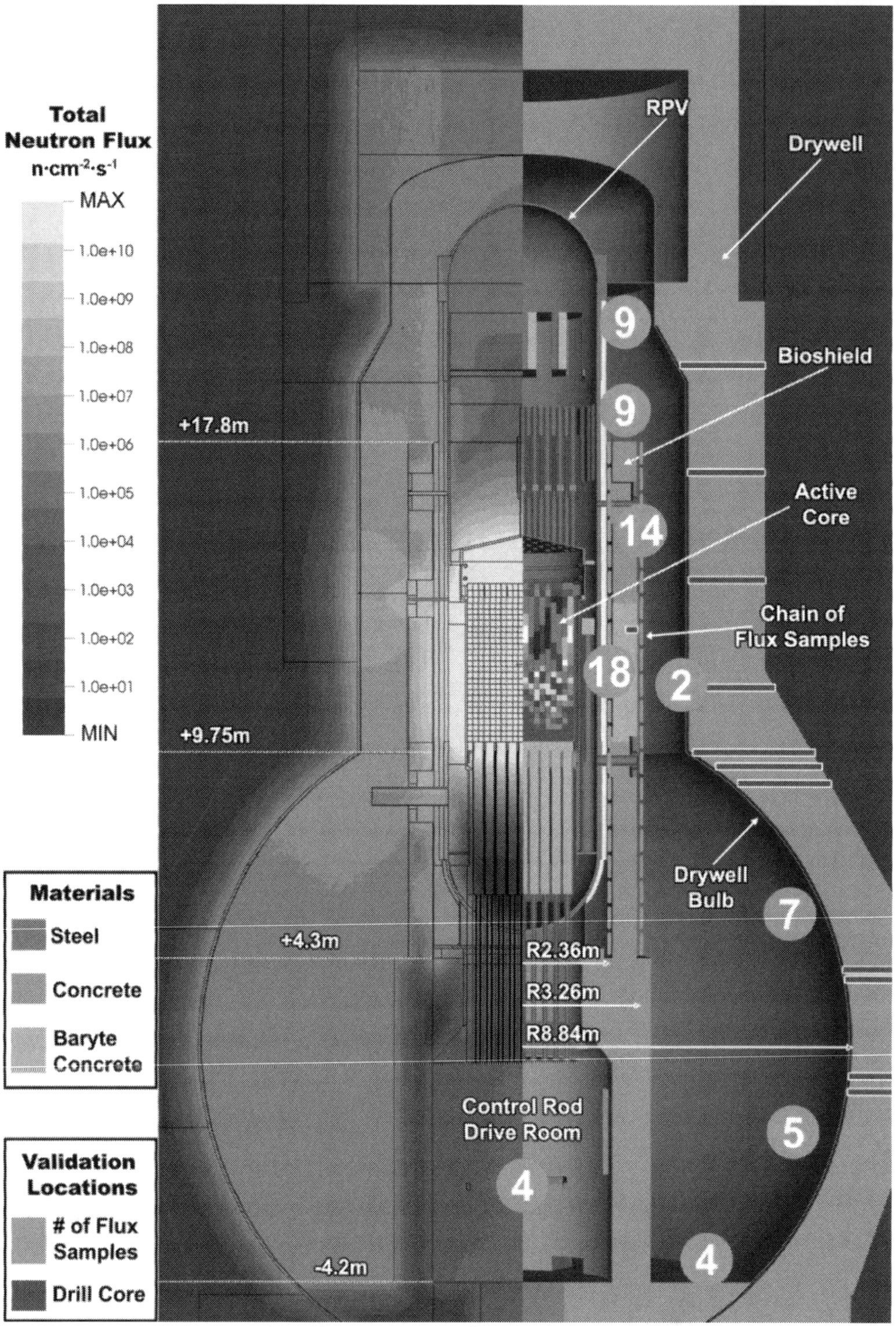

Figure 1. MCNP model of KKM, showing the total neutron flux distribution (left) and the modeled geometry (right), including a summary of the locations of the flux samples and the drill cores. The geometry is color-coded based on the material type. The number of flux samples in a particular area are indicated within orange circles. The locations of drill cores are indicated using purple rectangles. Key dimensions are provided as orientation for the sample positions defined in subsequent tables.

3.1 Methodology

The foil material is chosen based on the following three criteria. Since the sample is activated for one reactor cycle (in Switzerland typically approximately one year long), the chosen materials must have activation products with sufficiently long half-lives. Furthermore, the activation products need to be easily detectable, preferably with gamma spectrometry. Finally, the chosen foil materials should provide information about the energy distribution of the neutron flux. Based on these criteria, cobalt, silver, and nickel were chosen for AMAC validation. In subsequent validation campaigns, scandium was added to provide additional information about the epi-thermal energy range. Figure 2 shows the relevant cross sections, as well as an extruded cut view of a validation sample (foils inside an aluminum capsule). For both ^{59}Co and ^{109}Ag the majority of activation happens at the energy of the first peak (131 eV for ^{59}Co and 5.22 eV for ^{109}Ag, respectively). ^{58}Ni activation has a threshold of approximately 500 keV and majority of reactions occurring at even higher energies. As such, ^{58}Ni activation is representative of the fast neutron flux.

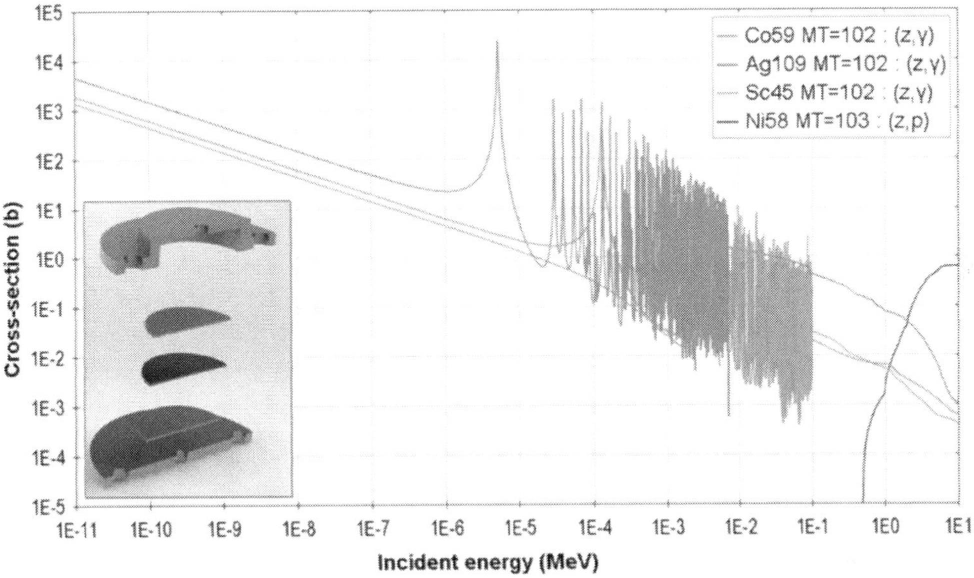

Figure 2. Cross sections of relevant foil activation reactions; bottom left: extruded cut view of a validation sample (foils inside a capsule).

The foil thickness is selected based on the expected neutron fluence (based on the past measurements and modeling experience with other plants), such that the activity of the resulting activation products is high enough for the spectrometric measurements, but at the same time low enough to be exempt from the regulatory control, as well as to avoid significant self-shielding.

The samples are either placed in locations which directly lie on key neutron streaming paths or in representative locations in rooms. They are attached to fixtures (such as pipes or handles) using an aluminum chain. The use of the more common steel chains is undesirable, due to neutron activation (since steel always contains cobalt impurities, producing ^{60}Co), which would necessitate the disposal of the chain as radioactive waste. In locations where only very low thermal flux is expected, the silver foil is omitted from the samples.

In order to compare the MCNP model simulation results with the foil measurements, the MCNP model is used to determine the reaction rates for the corresponding foil activation reactions. The variance reduction code ADVANTG [3] is used to accelerate these calculations. The calculated reaction rates are then combined with known cycle parameters (such as average power and cycle length) to determine the final activity of each corresponding foil. The agreement between the two is expressed as a ratio between the *calculated* and the *measured* value (C/M). Due to the small size of samples, substantial computational time is required for good statistics, given the large number of samples and the iterative nature of validation. Therefore, the Generic Sample Activation Model (GSAM) was developed [1]. Instead of implementing the small foils directly, a larger cell (typically a sphere of five centimeter diameter) is used to tally the neutron flux distribution. This result is then used to define the source in a simple MCNP model representing the sample inside the larger cell. This significantly reduces the computational time, albeit at the cost of introducing additional uncertainty, estimated to be in the order of 20% for a typical case [1].

3.2 Results

The C/M factors for all foils are shown in Figure 3, referring to the activities of the activation products of the three foils (^{60}Co, ^{110m}Ag, and ^{58}Co, respectively) plotted against the height (as indicated in Figure 1), since the axial trends dominate over radial and angular trends. The error bars are dominated by the aforementioned GSAM uncertainty (of approximately 20%), with reported activity measurement uncertainty typically around 2%.

The largest deviations occur in the bottom half of the drywell bulb (levels +1.0m, -2.7m, and -3.7m). Focusing on the C/M factors of the cobalt foils, as these are most representative of the neutron energy range leading to activation of cobalt and europium in concrete, we observe an overestimation of at least a factor of 2.5, which gradually increases with the decreasing height. At the height levels -2.72m and -3.72m the C/M factors range from 6.6 to 11.7, demonstrating a substantial overestimation in this area. The level -1.7m samples, corresponding to the control rod drive room, feature cobalt foil C/M factors between 2.5 and 3.6. For these sections, the deviations are primarily a result of the empty room assumption.

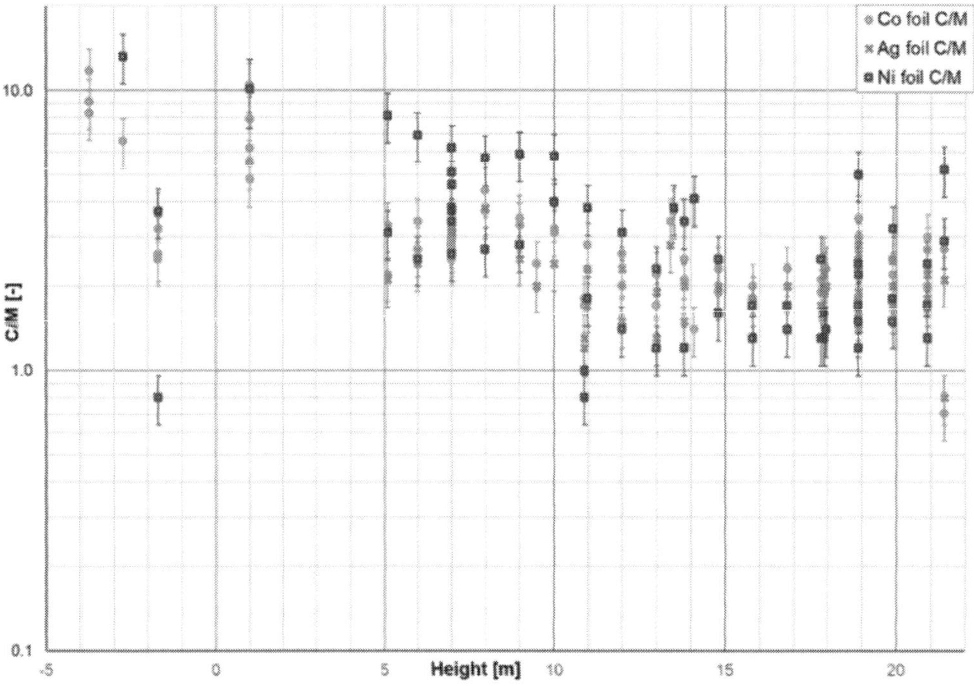

Figure 3. Flux Validation Results (C/M-values) plotted against reactor building height level.

https://doi.org/10.1051/epjconf/202327803001

4 Activation Validation through Gamma Spectrometry

The second technique used for the validation of AMAC models is based on the (mainly gamma, but sometimes also alpha and beta) spectrometry of the actual activated materials. For many components, such as the reactor internals, this can only be performed after the final shutdown of the plant, when the removal of material doesn't affect safety. However, limited sampling of the structures surrounding the RPV is possible even before the shutdown.

Three such campaigns are covered in this work. In August 2018, three samples were taken of the bioshield outer liner. Additionally, a 20-cm-long concrete drill core was extracted from the bioshield at +13.0m (just above the active core level) and divided into four equal sections. In August 2019, four drill cores were taken from the drywell wall at +16.6m, +14.7m, +11.8m, and +8.89, respectively. Following a detection of large overestimation at the lowest of these heights, a follow-up campaign was carried out in August 2020, extracting six drill cores at +9.6m, +9.4m, +4.3m, +4.2m, and (two cores at) +1.0m. The bioshield drill core is composed of baryte concrete, while all other drill cores are composed of normal concrete. No pieces of rebar reinforcement steel were detected in any of the drill cores. All drywell drill cores are shown in Figure 1, indicated by purple rectangles.

4.1 Methodology

Following the extraction, the drill cores were divided into smaller segments, with the range of depths corresponding to each segment recorded. These segments and the bioshield liner samples were then transferred to an on-site lab for gamma spectrometry evaluation.

Due to the high dose rate within the drywell, human access is limited. For this reason, only a few samples were extracted from the bioshield. Moreover, the drywell concrete drill cores were extracted from the outside of the drywell. Since none of the drill cores reached up to the inner liner of the drywell, this represents a potential source of uncertainty for the exact depth determination (being based on the outer drywell surface coordinates).

4.2 Results

The C/M factors for the aforementioned activation validation measurements are shown in Table 1 (for the bioshield) and Table 2 (for the drywell), respectively.

The uncertainties (not indicated in the tables) are primarily composed of the measurement uncertainties (in the order of 20%) and the calculation uncertainties—composed of uncertainties in location, impurity concentration, resolution of the energy group structure for activation calculations, and MCNP statistical uncertainty, altogether estimated to be in the order of 10%.

The C/M factors for the three samples of the bioshield liner show a significant variation, with the +17.8m sample being strongly overestimated and the other two samples being underestimated. These results also vary from the adjacent flux sample results. This variation in the bioshield liner sample results could be a product of variable cobalt impurity in the liner steel. Alternatively, it may be caused by the presence of shielding structures in the vicinity of the sample locations, which are not included in the MCNP model. However, due to the small overall number of bioshield liner samples, it is difficult to draw a general conclusion.

The concrete drill cores in the upper sections of the drywell (above the bulb) show a generally good agreement. The largest overestimation is again found in the drywell bulb section (as it was for the flux samples), and it grows with the decreasing height.

Table 1. Bioshield Activation Validation Results.

Bioshield liner

Height [m]	C/M ^{60}Co
7.1	0.5
13	0.5
17.8	19.3

Bioshield drill core 3, +13.0m

Depth [cm]	C/M ^{60}Co	C/M ^{152}Eu	C/M ^{133}Ba
0-5	1.3	1.8	1.2
5-10	1.6	2.0	1.3
10-15	2.6	3.5	1.8
15-20	2.7	3.3	1.4

Table 2. Drywell Activation Validation Results.

Drill core 4, +16.6m

Depth [cm]	C/M ^{60}Co	C/M ^{152}Eu
20.0	1.2	1.1
22.3	1.3	1.1
24.6	0.7	0.9
26.9	1.2	0.9
29.2	1.0	0.9
31.5	1.2	0.6
33.8	1.1	0.8
36.1	1.3	N/A
38.4	0.7	0.9
40.7	N/A	N/A
43.0	N/A	N/A
45.3	0.6	N/A

Drill core 3, +14.7m

Depth [cm]	C/M ^{60}Co	C/M ^{152}Eu
21.5	0.4	0.5
23.8	0.7	0.6
26.1	0.9	0.8
28.4	0.7	0.7
30.7	1.1	1.0
33.0	0.7	0.7
35.3	0.8	0.8
37.6	1.0	0.8
39.9	1.0	0.9
42.2	0.8	0.8
44.5	0.6	N/A
46.8	0.8	N/A
49.1	1.0	N/A

Drill core 2, +11.8m

Depth [cm]	C/M ^{60}Co	C/M ^{152}Eu
26.0	1.2	0.8
28.3	1.1	0.8
30.6	1.0	1.0
32.9	1.2	0.9
35.2	0.9	0.7
37.5	1.0	1.0
39.8	1.3	1.0
42.1	1.2	1.0
44.4	0.9	0.7
46.7	0.8	0.7
49.0	0.6	0.6
51.3	1.2	N/A
53.6	0.7	N/A

Drill core 11, +9.6m

Depth [cm]	C/M ^{152}Eu	C/M ^{60}Co
42.2	10.0	10.4
44.9	17.4	16.0
47.6	16.0	10.3
50.3	13.3	12.0
53.0	23.0	22.9
55.7	14.9	17.4
58.4	20.2	19.5
61.1	26.8	30.3
63.8	23.4	14.7
66.5	39.1	N/A
69.2	22.0	16.4
71.9	42.1	N/A
74.6	18.7	N/A

Drill core 12, +9.4m

Depth [cm]	C/M ^{152}Eu	C/M ^{60}Co
62.2	2.5	2.4
64.9	2.3	2.0
67.6	3.2	3.1
70.3	3.7	2.9
73.0	4.6	3.3
75.7	4.4	4.0
78.4	4.0	4.2
81.1	3.1	2.3
83.8	2.5	1.4
86.5	1.9	1.3
89.2	1.5	1.0
91.9	2.0	1.9
94.6	3.9	1.1
97.3	1.9	1.9
100.0	0.4	0.2

Drill core 1, +8.9m

Depth [cm]	C/M ^{60}Co	C/M ^{152}Eu
13.0	10.6	8.1
15.3	11.2	7.5
17.6	7.7	6.4
19.9	12.0	10.1
22.2	18.2	10.8
24.5	7.6	10.2
26.8	22.1	18.5
29.1	25.4	N/A
31.4	19.3	N/A

Table 2. Drywell Activation Validation Results (cont.)

Drill core 6, +4.3m

Depth [cm]	C/M ^{152}Eu	C/M ^{60}Co
12.2	13.9	16.3
14.9	11.6	13.2
17.6	11.4	10.0
20.3	10.8	11.9
23.0	8.5	7.2
25.7	9.9	11.2
28.4	8.6	6.6
31.1	11.8	7.3
33.8	15.6	N/A
36.5	24.0	N/A
39.2	11.2	N/A

Drill core 10, +4.2m

Depth [cm]	C/M ^{152}Eu	C/M ^{60}Co
10.2	6.0	7.1
12.9	5.0	9.2
15.6	9.1	5.9
18.3	2.3	8.6
21.0	8.4	7.5
23.7	7.0	8.4
26.4	9.5	10.9
29.1	8.6	N/A

Drill core 9, +1.0m

Depth [cm]	C/M ^{152}Eu	C/M ^{60}Co
6.2	23.6	17.9
8.9	23.6	20.8
11.6	19.8	22.7
14.3	14.2	10.7
17.0	19.3	13.5
19.7	13.4	7.1
22.4	7.9	7.4
25.1	5.0	5.2

Drill core 8, +1.0m

Depth [cm]	C/M ^{152}Eu	C/M ^{60}Co
10.2	12.3	16.8
12.9	14.9	13.6
15.6	14.8	17.0
18.3	13.2	8.3
21.0	13.4	13.1
23.7	12.8	N/A
26.4	16.2	N/A
29.1	11.7	11.1
31.8	24.0	N/A

5 Conclusion

In this work, the basis for a comprehensive validation of the KKM AMAC model simulation was presented, describing the results of two foil activation campaigns and three radiological measurement campaigns. The neutron flux and activation levels are found to be generally overestimated by the simulation, especially in the bottom half of the drywell bulb. These comparisons provide a first insight into the model's ability to predict neutron activation in the different regions of the KKM reactor building.

The future work will focus on uncertainty quantification, with a focus on the omitted room-filling structures (especially if shielding the validation samples from neutron radiation). Afterwards, the validation results will be used to define region-dependent scaling factors and thus enable the best estimate prediction of neutron activation for decommission planning.

The authors would like to acknowledge the support of Gerwin Spölgen (of BKW) and his contributions to organizing the KKM activation measurement campaigns.

References

1. V. Bykov, *High-Fidelity Determination of Nuclide Inventories for Radioactive Waste Disposal*, EPFL PhD Thesis (2019). DOI: 10.5075/epfl-thesis-8960.
2. C.J. Werner (editor), *MCNP Users Manual - Code Version 6.2*, LANL report LA-UR-17-29981 (2017).
3. S.W. Mosher et al. ADVANTG—An Automated Variance Reduction Parameter Generator. ORNL report TM-2013/416 (2021).

EPJ Web of Conferences **278**, 05001 (2023) https://doi.org/10.1051/epjconf/202327805001
ISRD 17

The Second EWGRD Round Robin: Inter Comparison of Gamma Spectrometry Measurements on Activation Dosimeters

Dean Thornton[a*], *James* Watson[a], *Hubert* Carcreff[b], *Christophe* Domergue[c], *Ana* Fernandes[d], *Pavel* Frajtag[e], *Pavlína* Haroková[f], *Daniel* Kis[g], *Vít* Klupák[h], *Vyacheslav* Kochkin[i], *Marie-Christine* Lépy[j], *Mladen* Mitev[k], *Rafał* Prokopowicz[l], *Vladimir* Radulović[m], *Ben* Straathof[n] *Jan* Wagemans[o], *Eva* Zsolnay[g]

[a] Jacobs Engineering, Gloucester, United Kingdom
[b] CEA, Service d'études des réacteurs et de mathématiques appliquées, Université Paris-Saclay, France
[c] CEA, DES, IRESNE, DER, Cadarache, F-13108 St-Paul-lez-Durance, France
[d] Instituto Superior Técnico, Universidade de Lisboa, Portugal
[e] École Polytechnique Fédérale de Lausanne, Lausanne, Switzerland
[f] Škoda JS, Plzen, Czech Republic
[g] Budapest University of Technology and Economics, Hungary
[h] Research Centre Řež, Czech Republic
[i] Kurchatov Institute, Moscow, Russia
[j] CEA, Laboratoire National Henri Becquerel, France
[k] Institute for Nuclear Research and Nuclear Energy of the Bulgarian Academy of Sciences, Bulgaria
[l] National Centre for Nuclear Research, Poland
[m] Jožef Stefan Institute, Slovenia
[n] NRG, Netherlands
[o] SCK-CEN, Belgium

Abstract. Following an initial Round Robin inter-comparison of gamma spectrometry measurements reported in 2014, this paper presents results from the first part of a second Round Robin inter-comparison commissioned by the European Working Group on Reactor Dosimetry in 2018. Measurements were performed by thirteen European organisations on a set of irradiated neutron activation detectors representative of those commonly used by the Reactor Dosimetry community to measure neutron fluence using gamma spectrometry methods. The radionuclides measured were ^{110m}Ag, ^{58}Co, ^{60}Co, ^{54}Mn, and ^{46}Sc. The purpose of the exercise was to demonstrate the level of consistency between participating organisations in blind tests of measurements. The samples used were disks of iron, nickel, titanium, and two standard alloys of aluminium, one of 1% cobalt and the other 1% silver. They were irradiated in the MARIA reactor operated by National Centre for Nuclear Research, Poland. Participants provided their results to an independent referee who collated and compared the data. The results are presented in an anonymised form together with discussion and conclusions which may be drawn from the exercise. Good agreement was obtained with standard deviations for individual measurements between 2.0% (^{54}Mn) and 5.6% (^{60}Co). Overall, the results obtained from the latter Round Robin show less consistency than those from the first Round Robin, which is attributed to the participation of a wider pool of organisations.

* Corresponding author: dean.thornton@jacobs.com

1 Introduction

Experimentally determined neutron activation reaction rates continue to perform an integral role within assessments of reactor dosimetry being routinely used to support, adjust and validate the multiplicity of calculational methods and assessments. It is therefore essential that such measurements are determined without systematic bias and with well-characterised uncertainties. Relying as they do upon gamma radiation spectrometry methods to determine neutron activation reaction rates *in-situ*, it is important for the reactor dosimetry community to establish the capabilities of the analysts and the facilities used. Benchmarks and blind tests are usually proposed for this purpose.

In 2014, the European Working Group on Reactor Dosimetry (EWGRD) reported the results of a Round Robin exercise [1] to compare neutron activation measurements performed by eight European facilities on five commonly used radionuclides: ^{110m}Ag, ^{60}Co, ^{54}Mn, ^{46}Sc and ^{94}Nb. Although the results reported were favourable, EWGRD members noted that the exercise would have been improved with a greater number of participants and the inclusion of other commonly used radionuclides, in particular ^{93m}Nb.

In response, the EWGRD has commissioned a second Round Robin inter-comparison involving fifteen facilities from eleven European countries this time including ^{93m}Nb in addition to the following commonly used gamma emitting nuclides: ^{54}Mn, ^{58}Co, ^{46}Sc, ^{60}Co and ^{110m}Ag. However, due to the unique challenges of measuring ^{93m}Nb, the Second EWGRD Round Robin was organised as two separate exercises, one for gamma-ray spectrometry of neutron activation products and the other for ^{93m}Nb. This paper describes the inter-comparison of gamma spectrometry measurements. The ^{93m}Nb inter-comparison is described in [2].

Thirteen organisations took part in the inter-comparison of gamma spectrometry measurements as follows:

- Budapest University of Technology and Economics, Hungary
- École Polytechnique Fédérale de Lausanne, Switzerland
- Instituto Superior Técnico, University of Lisbon, Portugal
- Institute for Nuclear Research and Nuclear Energy, Bulgaria
- National Centre for Nuclear Research, Poland
- Laboratoire National Henri Becquerel, CEA, France
- Škoda, Czech Republic
- CEA Cadarache, France
- Kurchatov Institute, Russia
- Jožef Stefan Institute, Slovenia
- Research Centre Řež, Czech Republic
- NRG, Netherlands
- SCK-CEN, Belgium

2 Preparation and Irradiation of Samples

Samples (in the form of activation foils) for the inter-comparison of activation measurements were prepared by SCK-CEN at their Mol facility. They included 25 cylindrical foils, five each of: iron (typically 35 mg); nickel (typically 7 mg); titanium (typically 52 mg); 1% silver aluminium alloy (typically 5 mg) and 1% cobalt aluminium alloy (typically 2 mg).

The irradiations took place on January 9th 2018 in the MARIA reactor operated by National Centre for Nuclear Research (NCBJ), Poland. In total 25 samples were irradiated, comprising five batches of five samples of each material. The irradiations were in a mixed thermal / fast spectrum at a nominal 1.2×10^{14} cm^{-2} s^{-1} and had the aim of achieving c. 50 kBq on each sample.

3 Protocol

Due to the challenges of radioactive consignment across national boundaries and the time it would take each laboratory to receive, measure and dispatch the samples, the approach taken for distribution differed from the first EWGRD Round Robin [1]. Instead of each participant measuring the same set of samples and forwarding to the next within a single chain, batches of five samples were circulated within subgroups of participants ensuring each participant measured one sample of each type. To act as control against variation between samples of the same material, one organisation (referred to as Org. 0) measured all of the samples prior to dispatching them to the other participants. One sample of each material was retained by Org. 0 as a precaution against loss.

On completion of its measurements, each participant was asked to forward its responses to an independent EWGRD member from a non-participating organisation, who collated the data and acted as referee. The data were treated anonymously, and the participants identified as Org. 0 to Org. 12, inclusive.

Participants were requested to collate sample mass measurements and specific activities, together with corresponding uncertainties within a pre-prepared template prescribing the units of measurement, number formats required and a reference time to which measured activities were to be decay-corrected (noon, 7th March 2018). Uncertainties (at one standard deviation, 1σ) were requested as the combination Type A, those that can be treated using statistical methods, and Type B, those than cannot.

Participants were also asked to provide:

- Sample mass measurements and uncertainties;
- Sample activities, and specific activities (activity divided by mass);
- A brief description of measurement technique(s) and equipment. including software;
- To identify any corrections applied or measures taken (e.g. dead time. self-absorption);
- Calibration, including any standard sources;
- Nuclear data: half-life values and photon emission yields assumed in the analysis;
- A short description of the treatment of uncertainties.

4 Sample Mass Measurements

Table 1 provides the properties and identification (ID) number of each sample, as well as the reported mass measurements and their associated uncertainties.

The mass data provided in Table 1 include the mean and standard deviation of the measurements carried out on each sample, as well as the maximum and minimum values obtained for each sample expressed as deviations from the mean in multiples of standard deviation. Measurements deviating ±2 standard deviations or more from the mean of each sample are shown in grey shade.

The table shows that mass measurements for each of the iron samples have produced consistent results within each subset of participants. The picture for the other dosimeter materials is less consistent. Poorer agreement is seen for the samples 1547 and 1548 (nickel), 1553 (titanium), 1557, 1558 (Al-Co) and 1562 (Al-Ag). All of these samples were measured by two subsets of organisations 0, 7, 8, 10, 11 (1547, 1557, 1562) and 0, 2, 3, 4 (1548, 1553, 1558). Of these Org. 2 produced a maximum value for each of the five samples it measured, and Organisation 3 produced minimum values for each of the five samples it measured. The results suggest the presence of systematic bias in some of the mass measurements, the effects of which will have been exacerbated by the relatively small masses of some of these samples.

Table 1 Summary of Sample Mass Measurements (see text for explanation of shading)

Sample	Type	Meas. by Org.	Mass Meas. 1	Mass Meas 2	Mass Meas 3	Mass Meas. 4	Mass Meas. 5	Mean	SD	Max (SD)	Min (SDs)
1541	Fe	0, 5, 6, 9	37.66 ±0.02%	37.66 ±0.03%	37.68 ±0.05%	37.68 ±0.02%	-	37.67	0.03%	0.34	-0.37
1542	Fe	0, 7, 8, 10, 11	34.99 ±0.03%	34.92 ±0.23%	34.92 ±0.06%	34.92 ±0.10%	34.90 ±0.29%	34.93	0.09%	2.00	-1.2
1543	Fe	0, 2, 3, 4	35.31 ±0.03%	35.40 ±0.57%	35.28 ±0.06%	35.33 ±0.51%	-	35.33	0.12%	2.39	-1.72
1544	Fe	0, 1, 12	37.69 ±0.03%	37.70 ±1.00%	37.68 ±0.01%	-	-	37.69	0.02%	0.31	-0.24
1545	Fe	0	37.64 ±0.02%	-	-	-	-	-	-	-	-
1546	Ni	0, 5, 6, 9	6.971 ±0.11%	6.972 ±0.13%	6.990 ±0.15%	6.971 ±0.06%	-	6.976	0.13%	0.10	-0.04
1547	Ni	0, 7, 8, 10, 11	6.977 ±0.11%	6.940 ±1.02%	6.940 ±0.15%	6.510 ±0.73%	7.000 ±1.45%	6.873	2.98%	0.98	-2.80
1548	Ni	0, 2, 3, 4	6.970 ±0.11%	7.100 ±1.44%	6.700 ±0.29%	6.980 ±0.72%	-	6.938	2.44%	1.24	-1.82
1549	Ni	0, 1, 12	6.954 ±0.12%	6.960 ±2.00%	6.953 ±0.04%	-	-	6.956	0.05%	0.03	-0.02
1550	Ni	0	6.975 ±0.11%	-	-	-	-	-	-	-	-
1551	Ti	0, 5, 6, 9	51.76 ±0.02%	51.75 ±0.04%	51.79 ±0.05%	51.75 ±0.01%	-	51.76	0.04%	0.53	-0.23
1552	Ti	0, 7, 8, 10, 11	51.82 ±0.02%	51.82 ±0.15%	51.80 ±0.04%	51.73 ±0.14%	51.80 ±0.19%	51.79	0.07%	0.46	-1.09
1553	Ti	0, 2, 3, 4	52.20 ±0.02%	52.30 ±0.19%	52.02 ±0.10%	52.21 ±0.52%	-	52.18	0.23%	2.06	-2.89
1554	Ti	0, 1, 12	52.22 ±0.02%	52.20 ±1.00%	52.22 ±0.01%	-	-	52.21	0.02%	0.15	-0.25
1555	Ti	0	51.87 ±0.02%	-	-	-	-	-	-	-	-
1556	Al-Co1%	0, 5, 6, 9	1.889 ±0.48%	1.889 ±0.42%	1.890 ±0.15%	1.888 ±0.20%	-	1.889	0.05%	0.01	-0.01
1557	Al-Co1%	0, 7, 8, 10, 11	1.912 ±0.47%	1.850 ±3.77%	1.930 ±1.08%	1.680 ±2.86%	1.900 ±5.39%	1.857	4.92%	0.84	-2.03
1558	Al-Co1%	0, 2, 3, 4	1.861 ±0.43%	2.000 ±5.43%	1.630 ±1.63%	1.870 ±1.63%	-	1.830	10.2%	1.97	-2.33
1559	Al-Co1%	0, 1, 12	1.917 ±0.47%	1.940 ±4.03%	1.917 ±0.16%	-	-	1.925	0.69%	0.17	-0.08
1560	Al-Co1%	0	1.845 ±0.49%	-	-	-	-	-	-	-	-
1561	Al-Ag1%	0, 5, 6, 9	5.235 ±0.15%	5.229 ±0.15%	5.239 ±0.15%	5.233 ±0.08%	-	5.234	0.08%	0.13	-0.13
1562	Al-Ag1%	0, 7, 8, 10, 11	5.171 ±0.15%	5.080 ±1.18%	5.110 ±0.39%	5.000 ±1.04%	5.100 ±1.96%	5.092	1.21%	2.17	-2.54
1563	Al-Ag1%	0, 2, 3, 4	5.224 0.17%	5.300 3.81%	5.210 ±0.38%	5.240 ±0.76%	-	5.244	0.76%	1.51	-0.89
1564	Al-Ag1%	0, 1, 12	5.228 0.17%	5.240 ±2.01%	5.226 ±0.06%	-	-	5.231	0.14%	0.23	-0.14
1565	Al-Ag1%	0	5.210 0.17%	-	-	-	-	-	-	-	-

Note: 1. Mass measurements 1 to 5 correspond respectively to the organisations identified in the third column.
2. The two columns on the right express deviations from the mean as multiples of standard deviation for highest and lowest measurements.

Table 2 Details of Specific Activity Measurements Provided by Each Participant (note: maximum and minimum values identified in grey shade, maximum values with white text)

		Org.0	Org.1	Org.2	Org.3	Org.4	Org.5	Org.6	Org.7	Org.8	Org.9	Org.10	Org.11	Org.12
^{54}Mn	Sp. Act. (Bqg^{-1})	6.59E+05	6.16E+05	6.48E+05	6.51E+05	6.57E+05	6.49E+05	6.57E+05	6.58E+05	6.44E+05	6.60E+05	6.76E+05	6.50E+05	6.58E+05
	Unc.	1.2%	2.4%	5.8%	1.1%	2.1%	1.2%	0.6%	0.9%	3.2%	1.7%	1.3%	1.5%	1.8%
	Δ to mean	1.1%	-5.5%	-0.7%	-0.3%	0.7%	-0.6%	0.7%	0.9%	-1.3%	1.1%	3.6%	-0.4%	0.8%
	Δ / σ_{meas}	0.5	-2.3	-0.1	-0.3	0.3	-0.5	1.2	1.0	-0.4	0.7	2.8	-0.3	0.5
	Δ / σ_{pop}	0.5	-2.7	-0.4	-0.1	0.3	-0.3	0.3	0.4	-0.6	0.6	1.7	-0.2	0.4
^{58}Co	Sp. Act. (Bqg^{-1})	6.22E+06	5.93E+06	5.88E+06	6.44E+06	6.28E+06	6.14E+06	6.26E+06	6.14E+06	6.00E+06	6.27E+06	6.74E+06	6.20E+06	6.12E+06
	Unc.	0.6%	2.9%	6.5%	1.2%	2.5%	1.2%	0.7%	1.4%	3.1%	1.7%	1.2%	3.2%	1.8%
	Δ to mean	0.3%	-4.5%	-5.1%	3.8%	1.3%	-1.1%	1.0%	-1.0%	-3.2%	1.1%	8.7%	0.0%	-1.3%
	Δ / σ_{meas}	0.4	-1.5	-0.8	3.1	0.5	-0.9	1.4	0.7	-1.0	0.6	7.4	0.0	-0.7
	Δ / σ_{pop}	0.1	-1.2	-1.4	1.1	0.4	-0.3	0.3	-0.3	-0.9	0.3	2.4	0.0	-0.4
^{46}Sc	Sp. Act. (Bqg^{-1})	5.09E+05	4.76E+05	4.74E+05	5.36E+05	4.89E+05	5.01E+05	5.06E+05	4.89E+05	4.77E+05	5.06E+05	4.99E+05	5.00E+05	4.94E+05
	Unc.	0.7%	2.8%	5.4%	1.5%	2.0%	1.0%	0.7%	0.8%	3.1%	1.7%	1.3%	2.0%	1.8%
	Δ to mean	2.5%	-4.2%	-4.5%	8.0%	-1.5%	0.8%	2.0%	-1.5%	-3.9%	1.9%	0.5%	0.7%	-0.5%
	Δ / σ_{meas}	3.7	-1.5	-0.8	5.2	-0.8	0.8	2.8	-1.9	-1.3	1.1	0.4	0.3	-0.3
	Δ / σ_{pop}	0.7	-1.2	-1.3	2.4	-0.5	0.2	0.6	-0.5	-1.2	0.6	0.1	0.2	-0.2
^{60}Co	Sp. Act. (Bqg^{-1})	2.23E+07	2.28E+07	2.13E+07	2.53E+07	2.30E+07	2.25E+07	2.26E+07	2.28E+07	2.14E+07	2.26E+07	2.67E+07	2.30E+07	2.26E+07
	Unc.	0.7%	4.3%	10.3%	2.2%	2.6%	1.1%	0.6%	3.8%	3.4%	1.7%	1.5%	4.3%	1.6%
	Δ to mean	-2.8%	-0.8%	-7.2%	9.9%	0.1%	-2.3%	-1.8%	-0.8%	-7.1%	-1.7%	16.2%	0.1%	-1.7%
	Δ / σ_{meas}	-3.9	-0.2	-0.7	4.4	0.0	-2.2	-2.9	-0.2	-2.1	-1.0	11.1	0.0	-1.0
	Δ / σ_{pop}	-0.4	-0.1	-1.1	1.6	0.0	-0.4	-0.3	-0.1	-1.1	-0.3	2.5	0.0	-0.3
^{110m}Ag	Sp. Act. (Bqg^{-1})	7.41E+06	7.11E+06	7.03E+06	6.53E+06	7.39E+06	7.42E+06	7.48E+06	7.33E+06	7.17E+06	7.51E+06	7.50E+06	7.10E+06	7.10E+06
	Unc.	0.8%	2.2%	9.8%	1.1%	2.3%	1.1%	0.7%	1.2%	4.0%	1.7%	1.5%	2.8%	1.8%
	Δ to mean	2.4%	-1.8%	-2.8%	-9.8%	2.1%	2.5%	3.3%	1.3%	-0.9%	3.8%	3.6%	-1.9%	-1.9%
	Δ / σ_{meas}	2.9	-0.8	-0.3	-8.8	0.9	2.3	4.4	1.1	-0.2	2.2	2.5	-0.7	-1.0
	Δ / σ_{pop}	0.6	-0.5	-0.7	-2.6	0.6	0.7	0.9	0.3	-0.2	1.0	1.0	-0.5	-0.5

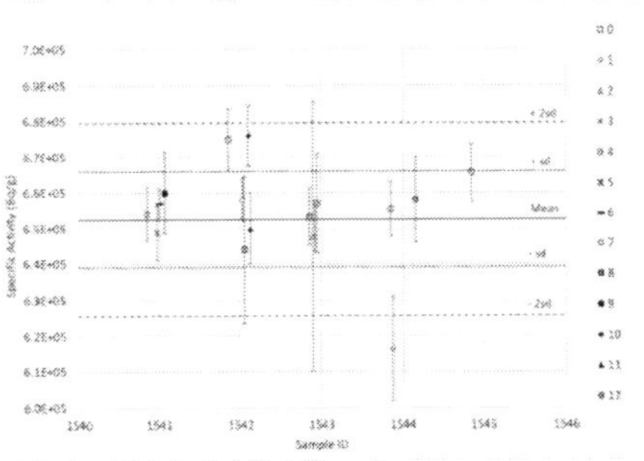

Fig. 1 Specific activity measurements obtained for ^{54}Mn

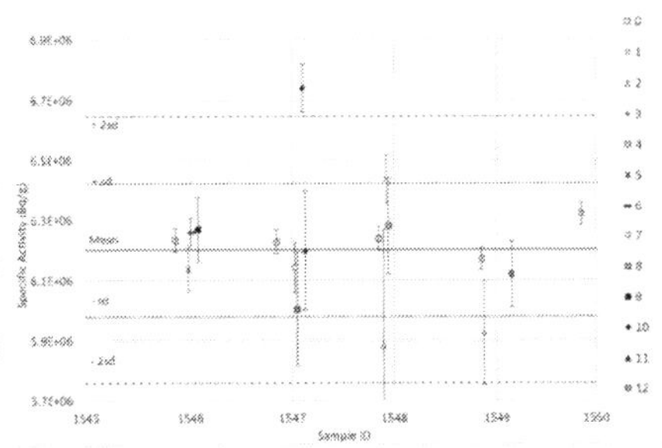

Fig. 2 Specific activity measurements obtained for ^{58}Co

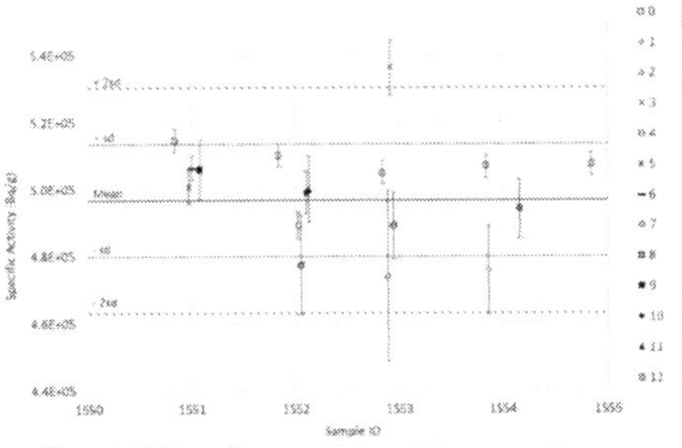

Fig. 3 Specific activity measurements obtained for ^{46}Sc

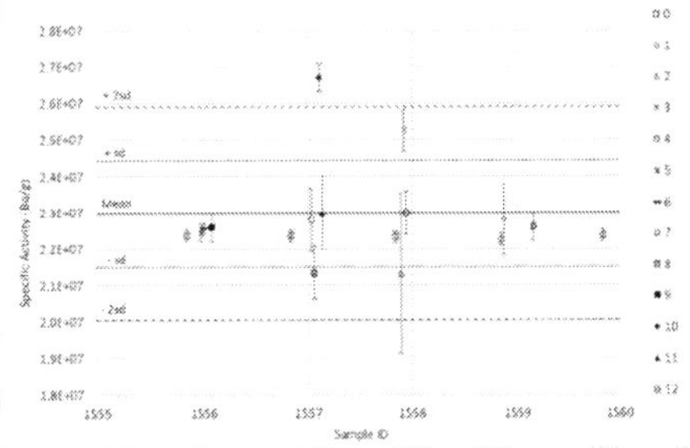

Fig. 4 Specific activity measurements obtained for ^{60}Co

 https://doi.org/10.1051/epjconf/202327805001

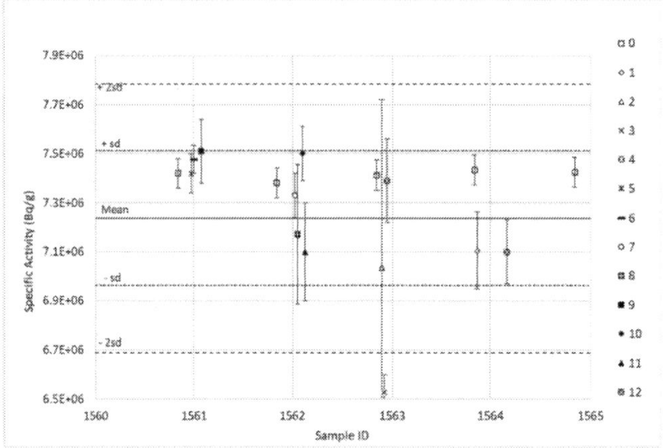

Fig. 5 Specific activity measurements obtained for ^{110m}Ag

5 Comparison of Specific Activity Measurements

Table 2 provides the measurement results for each participant and each nuclide, together with measurement uncertainties (σ_{meas}) and discrepancies (Δ) from the mean for each nuclide expressed as multiples of both σ_{meas} and the standard deviation of measurements obtained on for each nuclide (σ_{pop}). Cell shading identifies the maximum and minimum values obtained for each radionuclide. For Org. 0, the specific activities tabulated are the means of the measurements taken, excluding retained samples, and the corresponding σ_{meas} values are the means of individual measurement uncertainties corresponding to the data averaged.

In calculating the mean and standard deviation (σ_{pop}) for each radionuclide, Org. 0 data were treated as a single point represented by its mean (excluding retained samples). This was done to prevent Org. 0 having a dominant influence on the analysis.

Specific activity measurements for each of the sample materials are also shown in Figures 1 to 5. Each figure plots the specific activity measured by each participant against the sample number. Error bars show the measurement uncertainty provided by the participant at one standard deviation. The horizontal lines plotted show the mean and mean ±multiples of the population standard deviation (σ_{pop}) for each radionuclide. Note that the sample numbers plotted on the horizontal axis are artificially separated to aid identification. To determine whether any of the data should be considered as outliers, statistical tests have been performed using Grubbs' test [3].

For ^{54}Mn, the mean activity obtained was 6.53×10^5 Bq g^{-1} with a standard deviation (σ_{pop}) of 2.1%. Figure 1 shows good agreement between all participants with all but three inside $\pm\sigma_{pop}$ of the mean. One value (Org. 1) is at 2.7σ_{pop} below the mean, which can be considered an outlier with greater 5% confidence according to [3]. Taking into account the measurement uncertainties (plotted as error bars), all of the measurements appear consistent with the mean, and even the outlier appears reasonable. Org. 0's results display scatter comparable with both its measurement uncertainties and σ_{pop} but, other than that, suggest no significant variation between samples.

For ^{58}Co, the mean activity obtained was 6.20×10^6 Bq g^{-1} with a standard deviation (σ_{pop}) of 3.6%. Figure 2 shows a more scattered distribution with four measurements (Orgs. 1, 2, 3) outside $\pm\sigma_{pop}$ and one at 2.4σ_{pop}. above the mean (Org. 10). Again, taking into account measurement uncertainties, all measurements appear reconciled with the mean with the exception of Org. 10, which appears particularly discrepant and may be considered an outlier with 5% confidence according to [3]. With lower measurement uncertainties than the other participants, Org. 0's results sit close to the mean value, suggesting no sample-to-sample variation any greater than ±½ σ_{pop}.

The mean activity obtained for ^{46}Sc was 4.97×10^5 Bq g^{-1} with a standard deviation (σ_{pop}) of 3.4%. Figure 3 shows that all of the data are within $\pm\sigma_{pop}$ of the mean with the exception of Orgs. 1, 2 and 8, which are between -σ_{pop} and -2σ_{pop} below the mean. Org. 3 is 2.4 σ_{pop} above the mean and may be considered an outlier [3] though only just. Taking into account measurement uncertainties, all of the data appear reasonable compared to the mean except for Org. 3, the potential outlier, sitting 8% above the mean but with σ_{meas} of only 1.5%. Again, Org. 0 has noticeably lower measurement uncertainties than the other participants with no sample-to-sample variation greater than ~ ½ σ_{pop}. It is nevertheless interesting that for this nuclide, Org. 0 appears to be consistently greater than the mean by 2% on average and that its value for sample 1551 sits just over σ_{pop} above the mean.

For ^{60}Co, the mean specific activity obtained was 2.30×10^7 Bq g^{-1} with a standard deviation (σ_{pop}) of 6.4%. By some margin, this is the highest standard deviation of all five nuclides measured in the exercise. Figure 4

https://doi.org/10.1051/epjconf/202327805001

shows that, whilst nine of the participants' values are all close to the mean within $\pm\frac{1}{2}\,\sigma_{pop}$, four values are outside $\pm\sigma_{pop}$. Org. 10 is the most significant sitting 16% or $2.5\sigma_{pop}$ above the mean and considered an outlier according to [3]. At ±1.5%, Org. 10's σ_{meas} appears to have been greatly underestimated. Again Org. 0's measurements are provided with small measurement uncertainties and suggest no significant sample-to-sample variation although, this time they appear to be systematically 2% below the mean.

For ^{110m}Ag, the mean activity obtained was 7.24×10^{6} Bq g^{-1} with a standard deviation (σ_{pop}) of 3.8%. Figure 5 shows that, with the exception of Org.3, all data lie within $\pm\sigma_{pop}$ of the mean. Lying 2.6 σ_{pop} below the mean, Org. 3 is considered an outlier according to [3]. Measurements by Org. 0 suggest no discernible sample-to-sample variation, although it lies, on average, 2.0% above the mean.

6 Analysis and Discussion

6.1 Distribution of Discrepancies

Figure 6 presents histograms showing the distributions of measurement discrepancies relative to the mean for each nuclide expressed as multiples of σ_{meas} (histogram on the left) and σ_{pop} (histogram on the right). The data are presented so that the distributions are evident for both individual radionuclides, and the comparison as a whole. In addition to this, statistical testing has been carried out to test the distributions obtained for normality.

Compared to the standard deviations obtained for each nuclide, σ_{pop}, the overall picture is that the majority of measurements, 50 out of 65 (77%) are within $\pm\sigma_{pop}$ of their respective mean values and on top of this, an additional 10 (15%) lie within $\pm2\sigma_{pop}$. On this basis, the distribution appears normal, and despite the presence of 5 values at or close to $\pm3\sigma_{pop}$, the overall distribution successfully meets the criterion for 95% confidence in a chi-squared test for normality. The distributions of discrepancies for individual nuclides show a similar pattern of behaviour with all of them testing positive for normality with better than 95% confidence.

When compared to measurement uncertainties, σ_{meas}, the overall distribution is considerably more scattered and clearly not normally distributed, with two measurements less than $-3\sigma_{meas}$ and five greater than $+3\sigma_{meas}$. This is reflected in all of the individual nuclide data with the exception of ^{54}Mn, which produced no pronounced outliers and appears normally distributed, as supported by a chi-squared result.

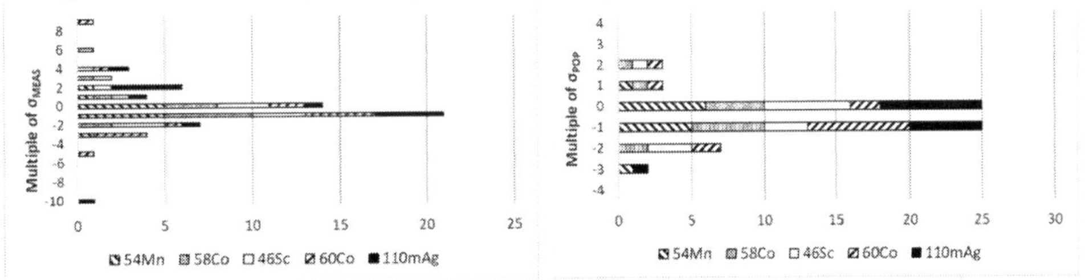

Fig. 6 Distributions of discrepancies from mean relative to measurement uncertainties, σ_{meas} (on the left) and relative to standard deviation, σ_{pop} (on the right).

The apparent normal distribution of discrepancies relative to the standard deviation for each nuclide, provides compelling evidence that the participants are producing consistent measurements. However, the contrast between the distribution of discrepancies relative to σ_{meas} and the distribution relative to σ_{pop} suggests that, for some participants at least, measurement uncertainties have been under-estimated, having failed to account for all of the factors leading to measurement errors, especially those of a systematic nature.

6.2 Compensation for Uncertainties in Mass Measurements

From the description of specific activity measurements in Section 5, it is apparent that some participants deviated systematically from the mean for a number of radionuclide measurements. For example, Orgs. 1, 2, 8 were amongst the lowest in at least three of the measurements, and Org 3 was high for ^{58}Co, ^{46}Sc and ^{60}Co. It is also evident from Table 1 that Orgs. 1, 2 and 3 tend to produce more of the extreme mass values in their cohort (subset of participants) than might be expected.

An obvious source of uncertainty is the measurement of sample mass. Because the masses of some of the samples are very small, most notably the nickel samples and the Al-Co and Al-Ag alloys, it is possible that for some participants, the mass of these samples was near the limit of what could be measured accurately with the equipment available. In such cases, the discrepancies between specific activity measurements could have arisen from the relatively straightforward measurement of mass rather than the considerably more technical challenge of gamma-radiation spectrometry.

To remove as far as possible the impact of mass measurement uncertainties from the comparison of specific activity measurements, each specific activity measurement has been adjusted by multiplying it by the factor $m_i^{org}/<m_i^{org}>$, where the numerator is the mass of sample *i*, as measured by organisation *Org* and the denominator is the mean of all mass measurements of sample *i*. The comparison of specific activities was then repeated using the specific activity data adjusted in this way. Note: this approach was taken despite the fact that participants were asked to provide sample activity data (Bq) in addition to specific activities (Bq g^{-1}). A comparison of the activity data supplied was not performed because the sample activities would have been influenced by the variation of masses between samples of the same type. Uncertainties on the mass-adjusted specific activities were also estimated by subtracting in quadrature the relative uncertainty on mass measurement from the relative uncertainty on the submitted specific activity. This makes the reasonable assumption that uncertainties on each participant's measurements of activity and mass are uncorrelated.

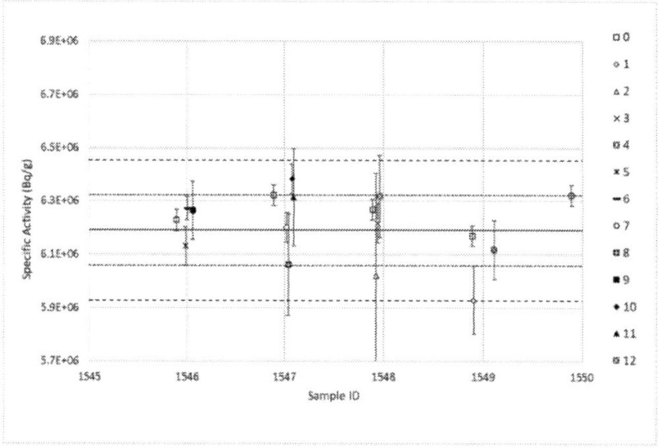

Fig. 7 Mass adjusted activity measurements for ^{58}Co

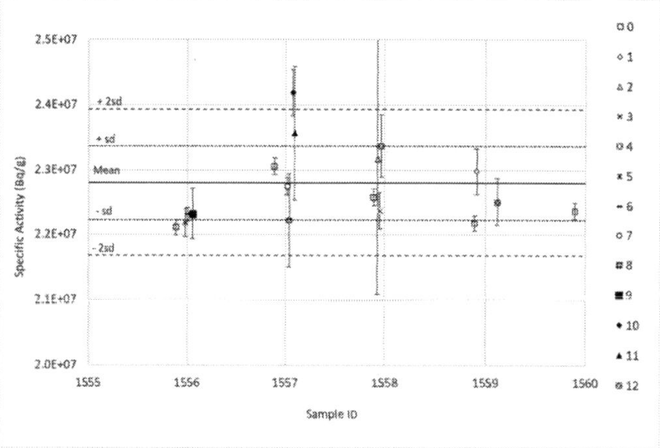

Fig. 8 Mass adjusted activity measurements for ^{60}Co

6.3 Comparison of Mass-Adjusted Specific Activity Measurements

Figures 7 and 8 show the impact of adjusting the mass data on respective ^{58}Co and ^{60}Co specific activity measurements. For ^{58}Co (Figure 7), the adjustments have produced a reduction in standard deviation (σ_{pop}) from 3.6% to 2.1% with no change in the mean. The adjustment has moved the measurement of Org 10 from an outlier at 2.4 σ_{pop} to being well within 2σ_{pop} of the mean. The value of Org 1's measurement has not changed significantly, but now falls just within -2σ_{pop} from the mean because of the reduction in σ_{pop}. Compared to the unadjusted data shown in Figure 2, the discrepancies evident in the adjusted data appear well represented by the revised measurement uncertainties.

In Figure 8, a similar picture emerges for the adjusted ^{60}Co measurements. Here there has been a marked reduction in σ_{pop} from 6.4% to 2.5% and a reduction in the range of data (maximum minus minimum) from 24% to 9%, as reflected in the revised range of the vertical axis compared with Figure 4. There has also been a 0.6% reduction in the mean. Analysis of the adjusted data in accordance with [3] shows that the Org. 10 value can no longer be considered an outlier.

Similar data for the other radionuclides are not shown because mass-adjustment produced little or no significant change. For ^{46}Sc, the adjustments produced a reduction in σ_{pop} from 3.4% to 3.3%, and for ^{110m}Ag a reduction from 3.8% to 3.7%. For ^{54}Mn, σ_{pop} remained unchanged at 2.1%.

It is important to note that the mass-adjusted data do not necessarily represent an improvement in every case. To illustrate this, Org. 0's ^{60}Co measurements have no significant sample-to-sample variation (as outlined in Section 5), but the adjustment creates artificial sample-to-sample variation, especially between samples 1556 and 1557. Such variation is an artifice created by the differences in individual organisation's mass measurements influencing the sample average.

The purpose of the mass adjustment is to determine the sensitivity of the inter-comparison to changing sample mass data to consensus values. Doing so provides an indication of the mutual consistency of the absolute activity measurements carried out by the participants on each batch of samples, with greatly reduced dependence on individual on sample mass measurements. It is emphasised that comparison of the as-submitted specific activities and the conclusions drawn from it forms the primary purpose of this work. Comparison of mass-adjusted data serves only to aid understanding of how deficiencies in individual mass measurements may have affected the results obtained.

6.4 Impact of Mass Adjustment on the Distribution of Discrepancies

Figure 9 presents histograms showing the distributions of discrepancies of the adjusted values relative to the mean for each nuclide expressed as multiples of the appropriate σ_{meas} and σ_{pop} values, as calculated from the adjusted data. Note: here σ_{meas} values have had contributions from mass uncertainties removed, as explained in Section 6.2.

Compared to the standard deviations of the adjusted data, σ_{pop} (right-hand histogram), chi-squared tests provide high confidence that the discrepancies remain normally distributed for each radionuclide as well as the data as a whole, i.e. unchanged from the as-submitted data, despite the reductions in σ_{pop}.

In comparison with Figure 6, the left-hand histogram shows that compared to measurement uncertainties (σ_{meas}), the distribution of discrepancies amongst the adjusted data has lost the three of the significant outliers at $\pm 6\sigma_{meas}$ but the single value at -10 σ_{meas} remains (Org. 3, ^{110m}Ag). Inspection of the data shows that although mass adjustment brought this point closer to the mean, there was a commensurate reduction in σ_{meas}. Other than that, the distribution appears considerably narrower. This is confirmed by a closer inspection of the data showing that 58 out of the 65 (89%) values lie within $\pm 3\sigma_{meas}$ suggesting that the distribution is still a little too scattered to be considered normal. Chi-squared analysis of the distributions for individual nuclides suggests that data for ^{54}Mn are normally distributed, but not for the other four radionuclides.

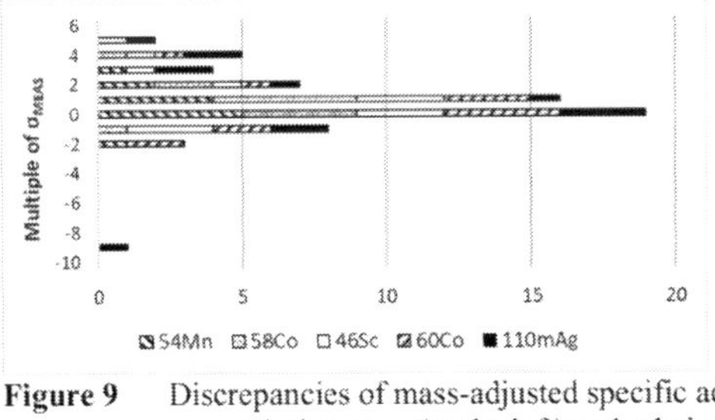

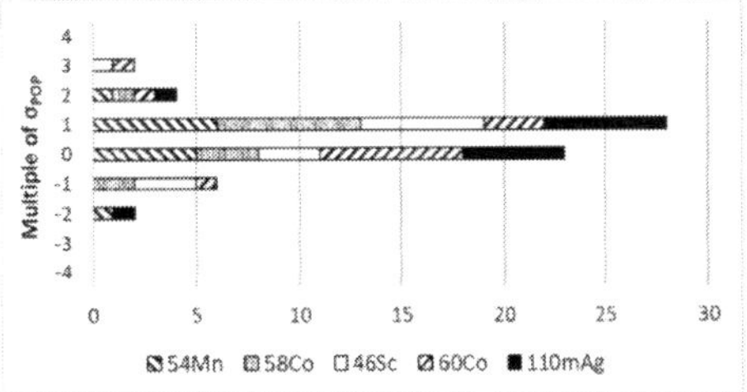

Figure 9 Discrepancies of mass-adjusted specific activities from mean relative to measurement uncertainties, σ_{meas} (to the left) and relative to standard deviation, σ_{pop} (to the right).

From these observations, it is concluded that the participants' estimates of the uncertainty on absolute activities are reasonable overall. In addition, the most significant outlying values obtained in the original comparison are a result of difficulties in measuring mass values, especially of the nickel and Al-Co samples. It remains a matter of debate as to why the ^{110m}Ag results did not show a similar improvement, given that the masses of the Al-Ag samples were also comparably low.

Table 3 Comparison of Normalised Measurements , $[A_i^{org}/<A_i^{org}>] - 1$

		Org.0	Org.1	Org.2	Org.3	Org.4	Org.5	Org.6
^{54}Mn	*Submitted*	1.1%	-5.5%	-0.7%	-0.3%	0.7%	-0.6%	0.7%
	Adjusted	1.1%	-5.5%	-0.5%	-0.4%	0.7%	-0.6%	0.7%
^{58}Co	*Submitted*	0.3%	-4.5%	-5.1%	3.8%	1.3%	-1.1%	1.0%
	Adjusted	0.9%	-4.2%	-2.7%	0.4%	2.1%	-1.0%	1.3%
^{46}Sc	*Submitted*	2.5%	-4.2%	-4.5%	8.0%	-1.5%	0.8%	2.0%
	Adjusted	2.5%	-4.2%	-4.3%	7.7%	-1.5%	0.8%	2.0%
^{60}Co	*Submitted*	-2.8%	-0.8%	-7.2%	9.9%	0.1%	-2.3%	-1.8%
	Adjusted	-1.4%	0.5%	1.3%	-2.2%	2.2%	-1.8%	-1.2%
^{110m}Ag	*Submitted*	2.4%	-1.8%	-2.8%	-9.8%	2.1%	2.5%	3.3%
	Adjusted	2.8%	-1.6%	-1.7%	-10.3%	2.1%	2.5%	3.5%
Mean	*Submitted*	**0.7%**	**-3.4%**	**-4.1%**	**2.3%**	**0.5%**	**-0.1%**	**1.0%**
	Adjusted	**1.2%**	**-3.0%**	**-1.6%**	**-1.0%**	**1.1%**	**0.0%**	**1.3%**
		Org.7	**Org.8**	**Org.9**	**Org.10**	**Org.11**	**Org.12**	**RMS**
^{54}Mn	*Submitted*	0.9%	-1.3%	1.1%	3.6%	-0.4%	0.8%	2.0%
	Adjusted	0.8%	-1.3%	1.2%	3.6%	-0.5%	0.8%	2.0%
^{58}Co	*Submitted*	-1.0%	-3.2%	1.1%	8.7%	0.0%	-1.3%	3.4%
	Adjusted	0.2%	-2.1%	1.2%	3.1%	2.0%	-1.2%	2.0%
^{46}Sc	*Submitted*	-1.5%	-3.9%	1.9%	0.5%	0.7%	-0.5%	3.2%
	Adjusted	-1.5%	-3.9%	1.9%	0.4%	0.7%	-0.5%	3.2%
^{60}Co	*Submitted*	-0.8%	-7.1%	-1.7%	16.2%	0.1%	-1.7%	6.1%
	Adjusted	-0.5%	-2.8%	-1.2%	5.8%	3.0%	-1.6%	2.4%
^{110m}Ag	*Submitted*	1.3%	-0.9%	3.8%	3.6%	-1.9%	-1.9%	3.6%
	Adjusted	1.1%	-0.5%	3.8%	1.8%	-1.7%	-1.9%	3.6%
Mean	*Submitted*	**-0.2%**	**-3.3%**	**1.2%**	**6.5%**	**-0.3%**	**-0.9%**	**2.6%**
	Adjusted	**0.0%**	**-2.1%**	**1.4%**	**2.9%**	**0.7%**	**-0.9%**	**1.6%**

6.5 Normalised Results

Table 3 shows a comparison of the discrepancies relative to the mean of each radionuclide carried out by each participant normalised as follows: $[A_i^{org}/<A_i^{org}>] - 1$ and expressed as percentages. A_i^{org} is the specific activity of radionuclide i measured by organisation org and $<A_i^{org}>$ is the mean of all measurements of radionuclide i. Because Org 0 measured every sample, an exception is made such that the value of A_i^{org} for Org 0 is the mean of the four samples measured by other organisations (i.e. excluding retained samples). Normalised in this way, the data permit the identification of systematic trends of under or over-prediction by any participating organisation.

The table presents the normalised discrepancies for both as-submitted data and the mass-adjusted data. Averages across all five radionuclides are presented for each participant as well as the RMS (root mean square) of all the normalised values for each radionuclide and the set as a whole. Comparing the RMS values for the as-submitted values with the mass-adjusted values provides a further way to quantify the impact of the mass adjustment.

From the data in Table 3, it can be seen that:

- As already noted, across the thirteen organisations, the averaged level of agreement is very good. The most discrepant as-submitted participants are Org 10 (+6.5%), Org 2 (-4.1%), Org 1 (-3.4%) and Org 8 (-3.3%)
- With mass-adjustment, the results for Org 10, Org 8 and Org 2 improve markedly to +3.0%, -2.1% -1.5%, although they, together with Org 1 (-3.0%) remain the four most discrepant participants.
- Changes in RMS values (as-submitted to mass-adjusted) are only significant for ^{58}Co and ^{60}Co, quantitatively confirming the observations made in Section 6.3 that these are the only two sets of measurements affected by inaccuracies in mass measurements.

7 Comparison with First EWGRD Round Robin

The first EWGRD Round Robin (RR1) intercomparison [1] was reported to the 15th International Symposium on Reactor Dosimetry in 2014 and described measurements carried out in 2012, some six years before the measurements described in this paper. Although there were important differences in approach between the two comparisons, it is informative to compare the outcomes of both. For this purpose, standard deviations have been chosen as a simple high-level measure of the mutual consistency between different organisations' measurements.

For each radionuclide Table 4 gives the mass, the total activity (specific activity × mass) and the standard deviation obtained from the measurements performed on each sample (RR1), or batch of nominally equivalent samples (2nd Round Robin, RR2). Although each intercomparison considered measurements of five different radionuclides, only the four radionuclides given in the table were common to both.

Table 4 Comparison of the Results from the Current Work with the First EWGRD Round Robin

	1st EWGRD Round Robin [1]			2nd EWGRD Round Robin (this work)					
	Mass (mg)	Act. (Bq)	SD	Mass (mg)	Act. (Bq)	Orig. SD	Adj. SD	Org. 0 SD	RR1 Orgs(1)
^{54}Mn	630.6	1.71E+04	3.10%	36.7	2.39E+04	2.1%	2.1%	1.4%	2.5%
	17.4	7.95E+05	1.60%						
^{46}Sc	2.9	1.23E+05	2.20%	52.0	2.58E+04	3.4%	3.3%	0.7%	2.8%
	3.1	9.39E+04	1.80%						
^{60}Co	3.7	2.04E+04	1.70%	1.87	4.30E+04	6.4%	2.5%	0.2%	2.6% (1.8%)
	12.7	5.07E+03	2.00%						
^{110m}Ag	1.8	1.10E+05	2.10%	5.20	3.76E+04	3.8%	3.7%	0.3%	2.3% (2.2%)
	20.4	2.08E+04	2.80%						
Note:	(1) Column 'RR1 Orgs' shows the standard deviations obtained from the subset of organisations that participated in RR1. Values in parenthesis are from mass-adjusted specific activities. Where only one value provided, mass-adjustment had no effect.								

As the table shows, in RR1 two samples of each radionuclide were measured, in most cases with notably different masses and activities. For the RR2, the table shows the average mass of the five samples provided for each nuclide, the average sample activity for each batch as well as the standard deviations obtained from the specific activity measurements performed for each batch, both for the as-submitted data and the mass-adjusted data. The second column from the right shows the standard deviation of the measurements carried out on all five samples in each batch by Org. 0, i.e. an indication of the variability within the control measurements

With the exception of the more massive ^{54}Mn sample measured in RR1, the original (i.e. unadjusted) standard deviations obtained in RR2 are all greater than those obtained in RR1. Even the reduced standard deviations resulting from the use of mass-adjusted data are higher than those obtained in RR1. This is an interesting result because it suggests poorer levels of consistency amongst the European dosimetry community in RR2, and it warrants some discussion.

As seen in the current work, measurements of mass can affect the accuracy of specific activation calculations, especially when small and/or less sensitive balances are used. However, it can be seen from Table 4, that whilst they vary, the masses measured in RR1 are comparable with those measured in RR2. Similarly, Table 4 shows that the activities of measured radionuclides on corresponding RR1 and RR2 samples are also comparable within an order of magnitude, suggesting that high or low activity levels are not the cause. It therefore seems very unlikely that the properties of the samples are the reason for the differences seen.

An important difference in approach between RR2 and RR1 was the measurement of batches of nominally equivalent samples by subsets of participants rather than all participants measuring all samples, as in RR1. Whilst proving expedient, especially for an inter-comparison with many participants, the RR2 approach creates the risk that sample-to-sample variability might influence the submitted measurements. The use of Org. 0 as a control for variability provided a means of assessing this. In Table 4, the second column from the right shows that for all except ^{54}Mn, sample-to-sample variability as indicated by the standard deviation in the Org. 0 results is less than 1% and is not sufficient to reconcile RR1 and RR2 results. Whilst the sample-to-sample variability indicated for ^{54}Mn would have an impact for this radionuclide, this is the one radionuclide where the two Round Robins appear more consistent.

Having discounted the explanations above, a final one is that the participants in the two inter-comparisons were different. Certainly, RR2 had more participants, thirteen compared to eight of whom only six participated in RR2. However, that would only affect the consistency of the inter-comparisons if the performance of the original RR1 participants were more consistent than those RR2 participants who had not taken part in RR1. To assess this, the final column in the table shows the standard deviation of the specific activities of RR2 measurements calculated for the subset of six organisations participating in both Round Robins. In cells where two values are shown, the value in parenthesis corresponds to the mass -adjusted data. Where only one value is provided, as-submitted and mass-adjusted values are the same.

For all except ^{54}Mn, the RR1 participants produce notably lower standard deviations, comparable with the those seen in the RR1 [1], albeit that for ^{60}Co it is the mass-adjusted data rather than the as-submitted that are consistent. For ^{54}Mn, the RR1 participants do not perform any better but, as already noted, the two Round Robins compare reasonably well for this radionuclide in any case.

In conclusion, therefore, it appears that the original RR1 participants performed more consistently within RR2 than those who did not. Whilst speculation on reasons for this are beyond the scope of this paper, it is possible that, having already participated in a Round Robin six years before, greater experience in the original participants is translated into more consistent performance.

8 Lesson Learned and Ideas for Improvement

Although Round Robin inter-comparisons are well established as a means of underwriting and validating tools and techniques in Reactor Dosimetry, there were aspects to this exercise that were novel and it is worth briefly discussing these and ideas for improving similar exercises carried out in future.

The measurement of batches of nominally equivalent samples by subsets of participants was a departure from the protocol followed in the original EWRGRD Round Robin. This proved advantageous because it allowed more organisations to participate without incurring time delays associated with measurements in series, which would have risked significant levels of decay for the shorted lived radionuclides. Whilst within batch variability (sample-to-sample) risks increasing the scatter between measurements, the use of a single organisation to measure each sample provides an effective control if that organisation performs measurements with high precision. As seen in this work, the control measurements indicated modest variability with 1.4% as the highest standard deviation in any of the five batches.

The standard of submissions received by the referee was high, with only a handful of typographical errors identified. This observation represents a significant improvement on RR1 where a number of errors were detected in the original submission. In the main part, this improvement can be attributed to the increased vigilance of the participants having learned lessons from RR1 [1]. However, the use of a document template clearly specifying the requirements of each participant is judged to have played a role in the improved performance.

Compared to RR1, where every participant measured the mass of each sample, the protocol adopted in RR2 has the disadvantage that relatively few (between three and five, excluding the retained samples) measurements of mass were performed for any one sample. Consequently, it has been difficult to make statistical inferences on the consistency of mass measurements. This is unfortunate because one of the most important observations from RR2 is that discrepancies in mass measurement have resulted in a number of outlying specific activity measurements. With greater confidence in the mass of each sample, it would have been possible to shift the focus away from sample mass measurements and more onto the activity measurements, i.e. the primary objective of the inter-comparison. To reduce the adverse impact of discrepant mass measurements on future inter-comparisons, the following could be adopted:

- Benchmark the masses in advance by obtaining independent high, precision measurements;
- Avoid small sample masses so that mass measurement uncertainties are minimised.

It is important too that participants provide absolute activity measurements together with their associated uncertainties. In the event that data have to be manipulated or adjusted, as done here, it is helpful to be able separate laboratory identified contributions such as mass.

9 Conclusions

The European Working Group on Reactor Dosimetry has undertaken a second Round Robin to compare gamma spectrometry measurements performed by thirteen European organisations on a set of irradiated dosimetry samples representative of those commonly used by the Reactor Dosimetry community to measure neutron fluence. The radionuclides measured were ^{110m}Ag, ^{58}Co, ^{60}Co, ^{54}Mn, and ^{46}Sc.

The results presented anonymously in this paper show that:

- The thirteen organisations have provided consistent measurements of specific activity for each of the radionuclides with standard deviations varying between 2.1% (^{54}Mn) to 6.4% (^{60}Co);
- Compared to the measurement uncertainties provided by the participants, 71% are within ±2 multiples of measurement uncertainty;
- Samples weighing c. 5 mg or less have resulted in larger than expected discrepancies in the results of some of the participants, a number of which have fed through as outliers in the comparison of specific activity data;

- Adjustment of (all) specific activity data to make them consistent with the average of the measured masses for each sample produced a significant improvement in overall agreement. The range of standard deviations obtained was reduced, ranging from to 2.1% (^{54}Mn and ^{58}Co) to 3.7% (^{46}Sc). Correspondingly, 90% of the mass-adjusted measurements lie within ±2 multiples of measurement uncertainty;
- Compared to the original European Working on Reactor Dosimetry Round Robin, the results of the latest inter-comparison show poorer consistency between organisations. However, analysis shows that for the subset of participants that took part in both Round Robins, similar levels of consistency were obtained.

Finally, it is concluded that the organisations participating in this inter-comparison have produced a consistent set of activity measurements with representative measurement uncertainties. However, the measurement of small masses could present a significant challenge with the potential to undermine the specific activity values which depend upon them.

References

1. D.A. Thornton, J. Wagemans, A.C. Fernandes, J.M. Girard, D.P. Kis, V. Klupák, R. Mutnuru, H. Philibert, G. Rousseau, J.P. Santos, T Serén, E. M. Zsolnay, *The EWGRD Round Robin Measurement Exercise*, paper submitted to the 15th International Symposium on Reactor Dosimetry, EPJ Web of Conferences **106,** 06005 (2016),), https://doi.org/10.1051/epjconf/201610606005
2. D.A. Thornton, C. Domergue, A. Fernandes, L. Gravier, P. Haroková, V. Klupák, V. Kochkin, M. Lépy, M. Mitev, R. Prokopowicz, V. Radulović, B. Straathof, J. Wagemans, E. Zsolnay, The Second EWGRD Round Robin: Inter-Comparison of ^{93m}Nb Measurements, paper submitted to the 17th international Symposium on Reactor Dosimetry, Lausanne, 2023
3. F.E Grubbs, Procedures for Detecting Outlying Observations in Samples, Technometrics, Vol. 11, No. 1, February 1969

Acknowledgements

The EWGRD would like to thank the participating authors and their organisations for their support in the completion of the exercise and the production of this paper. In particular, the authors would like to acknowledge the support of CEA, NCBJ and SCK-CEN whose commitment and initiative made the exercise possible.

The Second EWGRD Round Robin: Inter-Comparison of ^{93m}Nb Measurements

Dean Thornton[a*], *James* Watson[a], *Hubert* Carcreff[b], *Christophe* Domergue[c] *Tomasz* Dziel[d] *Ana* Fernandes[e] *Anita* Gerenyi [f] *Linda* Gravier [g] *Vít* Klupák[h] *Vyacheslav* Kochkin [i] *Marie-Christine* Lépy [j] *Mladen* Mitev [k] *Vladimir* Radulović [l] *Ben* Straathof [m] *Jan* Wagemans [n] *Eva* Zsolnay [f]

[a] Jacobs Engineering, Gloucester, United Kingdom
[b] CEA, Service d'études des réacteurs et de mathématiques appliquées, Université Paris-Saclay, France
[c] CEA, DES, IRESNE, DER, Cadarache, F-13108 St-Paul-lez-Durance, France
[d] National Centre for Nuclear Research, Radioisotope Centre POLATOM, Poland
[e] Instituto Superior Técnico, Universidade de Lisboa, Portugal
[f] Budapest University of Technology and Economics, Hungary
[g] Laboratoire d'Analyses Radiochimiques et Chimiques, CEA, France
[h] Research Centre Řež, Czech Republic
[i] Kurchatov Institute, Moscow, Russia
[j] Laboratoire National Henri Becquerel, CEA, France
[k] Institute for Nuclear Research and Nuclear Energy of the Bulgarian Academy of Sciences, Bulgaria
[l] Jožef Stefan Institute, Slovenia
[m] NRG, Netherlands
[n] SCK-CEN, Belgium

Abstract. Following on from an initial Round Robin inter-comparison of gamma spectrometry measurements reported in 2014, this paper presents the results of the second part of a further Round Robin inter-comparison commissioned in 2018 by the European Working Group on Reactor Dosimetry. The purpose of the exercise was to demonstrate the level of consistency between different organisations' measurements of the ^{93}Nb(n,n')^{93m}Nb reaction, which plays a key role in underwriting reactor dosimetry assessments. To achieve this, measurements of ^{93m}Nb activity were performed by twelve European organisations on six sets of near identical niobium samples, each having its own geometry and tantalum concentration. The samples were provided by CEA, France and irradiated in the MARIA reactor, operated by National Centre for Nuclear Research, Poland. Participants provided their results to an independent referee who collated and compared the data. The inter-comparison has demonstrated agreement to within standard deviations ranging from ±2.2% to ±7.9%, with a tendency for some organisations to measure elevated values. The results of the inter-comparison are presented in an anonymised form together with discussion and conclusions which may be drawn from the exercise.

1 Introduction

Over the last three decades, the neutron activation reaction ^{93}Nb(n,n')^{93m}Nb has come to play a key role in reactor dosimetry assessments because its energy dependence provides one of the best matches for fast neutron damage in steels. Reliant upon precision measurements of x-ray fluorescence, measurements of the ^{93m}Nb activation product demand specialist

* Corresponding author: dean.thornton@jacobs.com

techniques, sometimes requiring dissolution and hence destruction of some or all of the sample. Hitherto, the international reactor dosimetry community has not had an opportunity to benchmark measurements of this reaction.

This paper describes the results of an inter-comparison of ^{93m}Nb spectrometry measurements commissioned by the European Working Group on Reactor Dosimetry (EWGRD). It was organised as one of two separate inter-comparison exercises comprising the Second EWGRD Round Robin organised to follow the First EWGRD Round Robin reported to the 15th International Symposium on Reactor Dosimetry (ISRD) in 2014. [1]. The allied gamma activation inter-comparison is reported to the 17th International Symposium on Reactor Dosimetry (ISRD) in a separate paper [2].

Twelve organisations from ten European countries took part in the ^{93m}Nb inter-comparison, as follows:

- Budapest University of Technology and Economics, Hungary
- Instituto Superior Técnico, University of Lisbon, Portugal
- Institute for Nuclear Research and Nuclear Energy, Bulgaria
- POLATOM, Poland
- Laboratoire National Henri Becquerel, CEA, France
- Laboratoire d'Analyses Radiochimiques et Chimiques, France
- CEA Cadarache, France
- Kurchatov Institute, Russia
- Jožef Stefan Institute, Slovenia
- Research Centre Řež, Czech Republic
- NRG, Netherlands
- SCK-CEN, Belgium

2 Preparation and Irradiation of Samples

The niobium samples comprised 1 mg strip, and 8 mg disk geometries with tantalum impurity concentrations of 0.3 ppm, 4.0 ppm or 19.6 ppm for strips and 0.3 ppm and 4.0 ppm for disks. They were supplied by Laboratoire de Dosimétrie, de Capteurs et Instrumentation (LDCI), Cadarache, France. Each sample was given a name in the form Nb-n, where n is in integer in the range 1 to 61, not necessarily consecutive.

Tantalum impurities are known to affect x-ray fluorescence corrections in measurements of ^{93m}Nb activity. By providing samples with three different concentrations, the intention was to provide evidence for the impact, if any, of tantalum on the measurement results. Of the concentrations used: 0.3 ppm may be considered as pure; 4 ppm as a "standard" or typical concentration and 19.6 ppm considered as an impure sample.

The irradiations took place from 13th to 23rd October 2017 in the MARIA reactor operated by National Centre for Nuclear Research (NCBJ), Poland. In total 43 samples were provided for measurement, of which 38 were used in the inter-comparison.

3 Protocol

Because a significant proportion of the participating organisations rely on techniques requiring sample dissolution, it was not possible to carry out an inter-comparison along traditional lines in which each participant measures the same set of samples. Instead, distinct batches of near identical samples, each of which was subjected to the same irradiation conditions.

Using a non-destructive technique, one of the participating organisations was responsible for measuring each sample with high fidelity to provide a control for any variation between samples of the same batch. Once complete, subsets of samples were sent to each participant for measurement, with some samples retained as a precaution against loss.

Results were sent back to a non-participating member of the EWGRD who acted as referee by collating and analysing the data. The data were treated anonymously, and the participants identified as Org. A to Org. L, inclusive.

Participants were requested to record sample mass measurements and specific activities, together with corresponding uncertainties within a pre-prepared template, prescribing the reference time of the irradiation (against which specific activity data were to be decay-corrected), the units of measurement and number formats required. Uncertainties (at one standard deviation, 1σ) were requested as the combination Type A, i.e. those that can be treated using statistical methods, and Type B, i.e. those than cannot.

Participants were also asked to provide:

- Sample mass measurements and uncertainties;
- Sample activities, and specific activities (activity divided by mass);
- A brief description of measurement technique(s) and equipment, including any relevant software;
- Identify any corrections applied or measures taken (e.g. for dead time, self-absorption, fluorescence correction);
- Calibration. including any standard sources;
- Nuclear data used: i.e. what libraries were used or, where allocated separately, the values of half-lives and photon emission yields assumed in the analysis;
- A short description of the treatment of uncertainties.

4 Sample Mass Measurements

Table 1 shows the identification (ID) number, properties, mass measurements and associated uncertainties for each of the samples included in the inter-comparison, as well as the organisations measuring each sample. Sample properties are defined in terms of geometry, mass and approximate tantalum (Ta) concentration, which constituted six batches of samples (B1 to B6), each of which subject to the same irradiation conditions.

Organisation L was the one responsible for measuring all samples. Although the samples retained by Org. L were also measured, they are not included in Table 1 and have not played any role in the inter-comparison, other than as control for sample-to-sample variation.

On the whole, the mass measurements given in the table show excellent agreement between each pair of organisations. However, for the reasons given below, there are limitations on what can be inferred from the data:

- With only two values per sample, it is difficult to make statistical inferences on the accuracy of any individual's mass measurements as done in the accompanying EWGRD inter-comparisons on gamma spectrometry measurements [1] and [2].
- Org. K used dissolution techniques and only weighed a fraction of specimens Nb-41 and Nb-42. The reported mass values are substantially lower than the corresponding values reported by Org. L.
- Several organisations reported that samples had not arrived intact (Nb-4, Nb-31, Nb-45, Nb-46, Nb-51 and Nb-52), some in multiple pieces. Consequently, there is uncertainty that the recipient organisations measured all material weighed by Org. L and for these samples, the comparison of mass measurements may not be relevant, as shown by the systematically lower value of the second measurement in some circumstances.

Of the 38 pairs of data, 30 make no reference to sample fragmentation (third bullet) nor division of the sample for dissolution (second bullet). Within this subset, analysis of the data presented in Table 1 showed that 26 agreed within ±3 standard deviations(σ_m) of each other and 21 pairs show agreement within ±1%. Of the four outside ±3 σ_m three were measured by Org. E (Nb-7, Nb-18 and Nb-35) suggesting issues with mass measurements and/or

https://doi.org/10.1051/epjconf/202327805002

Table 1 Details of Samples and Mass Measurements

Specimen[(1)]	Description	Meas. by	**Value 1** (mg)	**Value 2** (mg)[(4)]
Nb-1	Batch 1(B.1) 1 mg strip, 0.3 ppm	L, G	0.882±0.008	0.887±0.007
Nb-2		L, J	0.912±0.008	0.915±0.008
Nb-3		L, B	0.957±0.009	1.000±0.100
Nb-4 [(2)]		L, H	0.977±0.008	(0.623±0.022)
Nb-5		L, I	0.824±0.008	0.820±0.020
Nb-6		L, F	0.883±0.008	0.884±0.001
Nb-7		L, E	0.925±0.008	0.880±0.002
Nb-10	Batch 2 (B.2) 1 mg strip, 4 ppm	L, G	1.087±0.008	1.084±0.007
Nb-11		L, J	1.165±0.008	1.165±0.008
Nb-12		L, B	0.980±0.008	1.000±0.100
Nb-13		L, H	1.202±0.009	1.209±0.025
Nb-14		L, I	1.261±0.008	1.260±0.020
Nb-15		L, F	1.087±0.009	1.087±0.002
Nb-18		L, E	1.380±0.008	1.170±0.002
Nb-19	Batch 3 (B.3) 1 mg strip, 19.6 ppm	L, G	1.256±0.008	1.255±0.007
Nb-20		L, J	0.871±0.008	0.858±0.008
Nb-21		L, B	1.025±0.008	1.000±0.100
Nb-22		L, H	1.193±0.009	1.188±0.025
Nb-23		L, I	1.068±0.008	0.980±0.020
Nb-24		L, F	0.738±0.009	0.738±0.001
Nb-25		L, E	1.042±0.008	1.020±0.002

Specimen[(1)]	Description	Meas. by	**Value 1** (mg)	**Value 2** (mg)[(4)]
Nb-28	Batch 4 (B.4) 8 mg disk, 0.3 ppm	L, G	9.103±0.008	9.102±0.009
Nb-30		L, D	8.615±0.009	8.610±0.020
Nb-31 [(2)]		L, E	9.062±0.009	(8.170±0.002)
Nb-41 [(3)]	Batch 5 (:B.5) 8 mg disk, 0.3 ppm	L, K	9.048±0.009	(1.485±0.045)
Nb-42 [(3)]		L, K	8.805±0.009	(1.290±0.039)
Nb-45 [(2)]		L, C	8.812±0.009	(7.636±0.003)
Nb-46 [(2)]		L, C	9.053±0.008	(8.952±0.004)
Nb-51 [(2)]		L, A	9.144±0.010	(2.515±0.003)
Nb-52 [(2)]		L, A	9.057±0.008	(2.866±0.003)
Nb-33	Batch 6 (B.6) 8 mg disk, 4 ppm	L, G	8.783±0.008	8.781±0.009
Nb-34		L, B	9.320±0.009	9.300±0.100
Nb-35		L, E	8.819±0.009	9.130±0.002
Nb-36		L, C	9.503±0.008	9.501±0.004
Nb-37		L, C	9.430±0.008	9.427±0.004
Nb-40		L, D	9.065±0.008	9.020±0.020
Nb-61		L, A	8.977±0.008	8.975±0.003
Nb-62		L, A	9.011±0.009	9.010±0.003

Notes:
(1) Only samples measured by two organisations are shown.
(2) Found to be broken into multiple fragments on receipt.
(3) Only fraction of sample measured.
(4) Mass measurements performed on incomplete samples are shown in parenthesis because their comparison with Org. L is not valid.

Table 2 Details of Specific Activity Measurements Provided by Each Participant

		Org L[(2)]	Org K[(3)]	Org J	Org I	Org H	Org G	Org F	Org E	Org D	Org C[(3)]	Org B[(1)]	Org A[(3)]	Mean [(1)] σ_{pop}
Technique: solid (S) or dissolution (D)		S	D	D	D	S	S	D	D	S	D	S	D	n/a
Batch 1 1mg strip 0.3ppm	Meas. Act.	2.57E+07		2.55E+07	2.75E+07	2.79E+07	2.60E+07	2.81E+07	3.10E+07			*5.00E+07*		2.74E+07
	$\sigma_{meas.}$	2.6%		6.7%	3.3%	4.7%	3.5%	3.9%	0.9%			*20%*		7.0%
	Δ to mean	-6.2%		-6.9%	0.3%	1.9%	-4.9%	2.7%	13.2%			*82.6%*		
Batch 2 1mg strip 4 ppm	Meas. Act.	2.59E+07		2.59E+07	2.74E+07	2.81E+07	2.56E+07	2.76E+07	3.50E+07			*5.00E+07*		2.68E+07
	$\sigma_{meas.}$	2.6%		7.3%	2.7%	3.9%	3.5%	3.9%	1.1%			*20%*		4.0%
	Δ to mean	-3.0%		-3.2%	2.3%	5.0%	-4.4%	3.3%	30.8%			*86.9%*		
Batch 3 1mg strip 19.6 ppm	Meas. Act.	1.82E+07		1.70E+07	1.77E+07	2.07E+07	1.76E+07	1.83E+07	2.01E+07			*3.30E+07*		1.85E+07
	$\sigma_{meas.}$	2.6%		7.1%	3.1%	3.9%	3.6%	4.0%	1.1%			*24%*		7.4%
	Δ to mean	-1.5%		-8.2%	-4.6%	11.8%	-5.0%	-1.1%	8.6%			*78.3%*		
Batch 4 8mg disk 0.3ppm	Meas. Act.	2.57E+07					2.60E+07		2.95E+07	2.99E+07				2.78E+07
	$\sigma_{meas.}$	2.6%					3.4%		1.0%	11.5%				7.9%
	Δ to mean	-7.3%					-6.3%		6.2%	7.4%				
Batch 5 8mg disk 0.3 ppm	Meas. Act.	1.74E+07	1.78E+07								1.73E+07		1.78E+07	1.76E+07
	$\sigma_{meas.}$	2.6%	5.1%								4.0%		2.8%	2.2%
	Δ to mean	-1.1%	1.0%								-1.7%		1.3%	
Batch 6 8mg disk 4 ppm	Meas. Act.	8.08E+06					7.97E+06		8.83E+06	9.35E+06	7.92E+06		8.15E+06	8.30E+06
	$\sigma_{meas.}$	2.6%					3.4%		1.2%	11.4%	4.0%		2.8%	6.2%
	Δ to mean	-2.6%					-4.0%		6.4%	12.7%	-4.5%		-1.8%	

Notes:
(1) Org. B (values in italic) not included in calculation of mean and σ_{pop}.
(2) For Org. L, values given are mean of measurements on all samples in batch.
(3) Each of the values shown for Org. K, Org C, and Org. A are the average of measurements performed on two samples in each group.

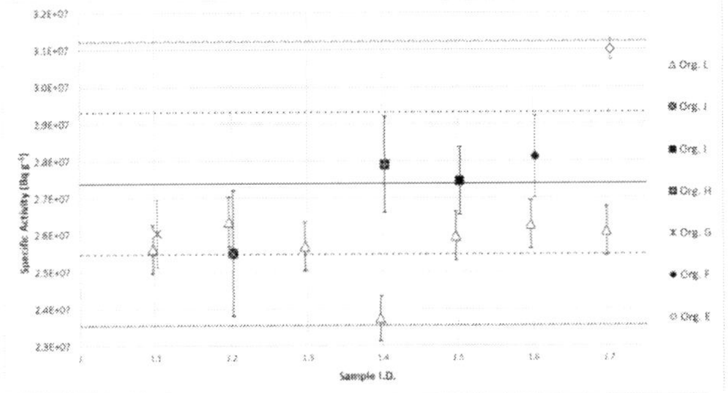

Fig. 1 ^{93m}Nb Specific activity measurements obtained for B.1, 1mg strip with 0.3ppm Ta

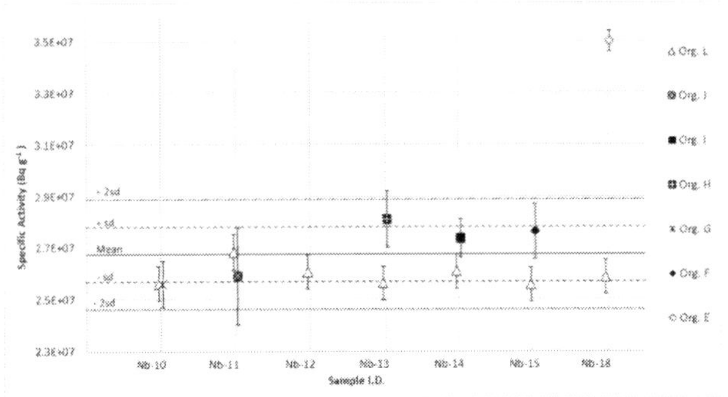

Fig. 2 ^{93m}Nb Specific activity measurements obtained for B.2,1mg strip with 4 ppm Ta

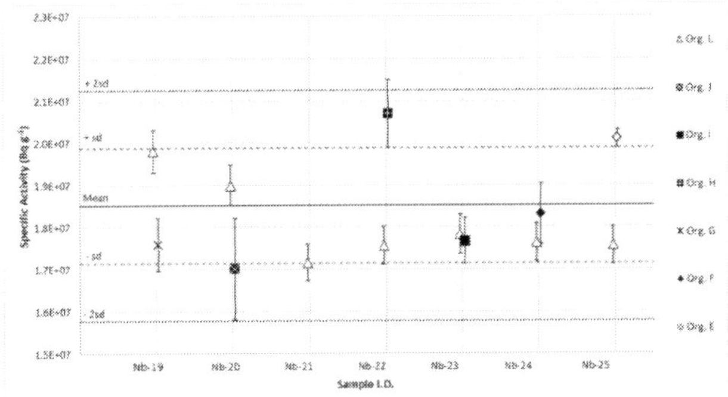

Fig. 3 ^{93m}Nb Specific activity measurements obtained for B.3,1mg strip with 19.6 ppm Ta

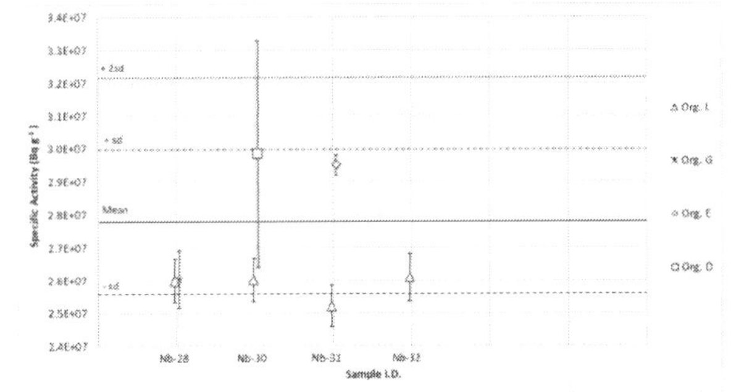

Fig. 4 ^{93m}Nb Specific activity measurements obtained for B.4, 8mg disk with 0.3 ppm Ta

https://doi.org/10.1051/epjconf/202327805002

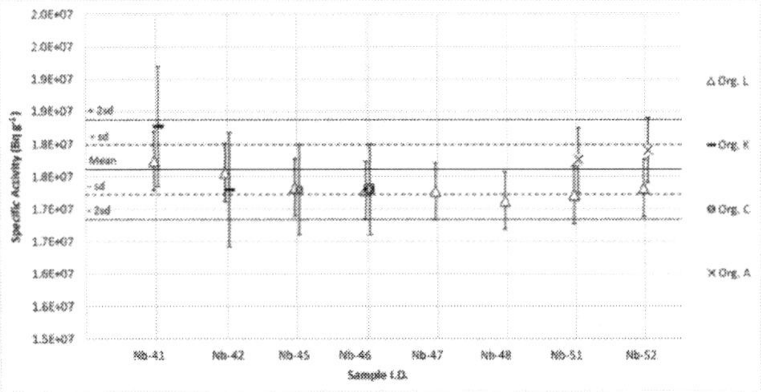

Fig. 5 ^{93m}Nb Specific activity measurements obtained for B.5, 8mg disk with 0.3 ppm Ta

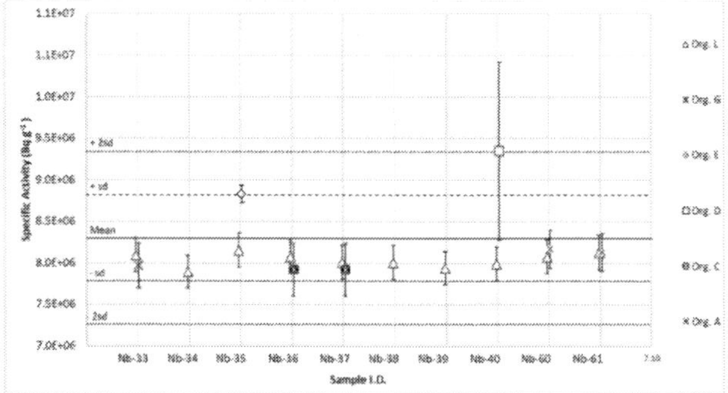

Fig. 6 ^{93m}Nb Specific activity measurements obtained for B.6, 8mg disk with 4 ppm Ta. Note, Nb-33 was measured by Org. G

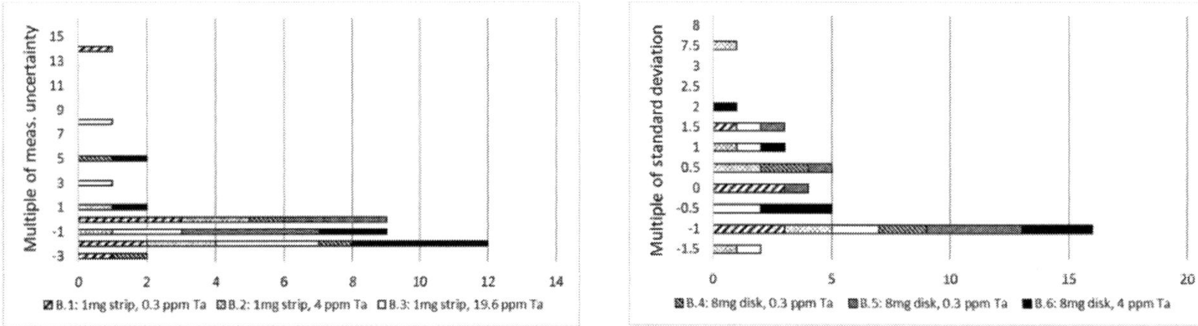

Fig. 7 Distributions of discrepancies from mean relative to measurement uncertainties, σ_{meas} (on left) and relative to standard deviation, σ_{pop} (on right). The legend applies to both graphs.

inadequate allocation of uncertainties. However, only Nb-18 and Nb-23 are more than ±5% different: Nb-18 (Org.L and Org E) with a difference of +17.9% and Nb-23 (Org. L and Org. I) a difference of +9.0%. If the data for these two samples represent a genuine difference in the measured masses, they are large enough to significantly affect the corresponding comparison of specific activity measurements (see Sections 6 and 6.1).

5 Specific Activity Measurements

Individual specific activity measurements of samples from each of the six batches are compared in Figures 1 to 6, which provide a graphical comparison of the specific activities obtained for each batch. Each figure shows the specific activity measured by each participant against the sample number. Error bars show the individual measurement uncertainty provided by the participant at one standard deviation. The horizontal lines plotted show the mean and multiples of the population standard deviation (σ_{pop}) for the batch, given in Table 2 as described below.

Table 2 compares the ^{93m}Nb specific activity results for each batch from each of the participating organisations. Note: except for Organisation L, participants measured a subset of batches, hence the empty cells in the table. For each organisation, the table provides the measurement technique used; the specific activity obtained and its associated measurement uncertainty (σ_{meas}); and the deviation (Δ) of the measured activity from the mean for the batch, expressed as a percentage. In order to provide a single representative value for each organisation's batch measurements averages were taken as follows:

- For Org. L, the data presented represent the mean of all samples within the batch (excluding retained samples);
- For Orgs. A, C and K, because each received pairs of samples in the same batches, the data shown in Table 2 for these participants represent the mean of a pair of samples.

In its righthand column, Table 2 shows the mean specific activity obtained for each batch, together with the standard deviation of the data for that batch, σ_{pop}. In calculating the mean and standard deviation, the Org. L data were treated as a single point represented by its mean (excluding retained samples). This was done to prevent Org. L having a dominant influence on the analysis.

From the data provided in Table 2, it is evident that the specific activity data provided by Org. B are significantly greater than their respective batch means, with substantially greater uncertainties for all samples measured. This was discussed with the correspondent who confirmed that ^{93m}Nb measurement is a technique in development at that organisation, for which the inter-comparison provided a useful exercise. With the participant's consent, it was deemed that Org. B's data were not representative of the data as a whole and would not be included in the global analysis (mean and standard deviations) nor the intercomparison of the data. These data are not shown in Figures 1 to 6.

The available measurements from all individual samples are represented in Figures 1 to 6, including those performed by Org. L on the retained samples, and the pairs of data from Orgs. A, C and K which were averaged in Table 2.

6 Discussion and Analysis

For the first batch of samples (1 mg strip with 0.3 ppm Ta), Figure 1 shows a generally good level of agreement, with all measurements lying within $\pm 2\sigma_{pop}$ of the mean value for this batch, and only two values outside the standard deviation, σ_{pop}, of ±7%. Compared to the population of data plotted in Figure 1, the position of the mean appears higher than might be expected from the disposition of data. However, it is important to recognise that seven of the data points lying below it are from Org. L, of which only the average value contributes to the mean and σ_{pop}..

With the exception of Nb-4, measurements performed by Org. L on all seven samples show consistency within their associated measurement uncertainties of 2.6%, i.e. no evidence of sample-to-sample variation. Whilst they are both within ±2 σ_{pop}, Org. E's measurement of Nb-7 and Org. L's value for Nb-4 are clearly the most outlying. The presence of a more discrepant value in the Org. L data does not adversely affect the mean value and σ_{pop} for this batch because it only represents one value out of the six used to calculate the average measurement of Org. L. By contrast, Org. E's outlying measurement of Nb-7 does affect the σ_{pop} mean and standard deviation for this batch. Excluding this outlier would reduce the mean from 2.74×10^{7} Bq cm^{-2} to 2.68×10^{7} Bq cm^{-2}, a reduction of 2.2% and a reduction in from 7% to 4%. In part, at least, Org. E's measurement of Nb-7 is likely to have been affected by the mass measurement which was 4.9% lower than that obtained by Org. L.

Figure 2 shows a similar pattern for the measurements performed on the second batch, 1 mg strip with 4 ppm Ta. The measurements performed by Org. L on each of these samples are mutually consistent within measurement uncertainties (σ_{meas}) of ±2.6%, indicating no sample-to-sample variation. Although the agreement between organisations is generally good, there is a potentially significant outlier on the measurement of Nb-18 by Org. E. Being some 30% greater than all the other measurements performed on this batch, Org. E's measurement of Nb-18 is considered anomalous and is associated with the 17% difference in σ_{pop}. mass measurements reported in Table 1. As a result, it is excluded from the calculated mean and σ_{pop} for this batch given in Table 2 and represented in Figure 2. With this point rejected, all but five data lie within $\pm\sigma_{pop}$ (4%) and those that don't lie comfortably within ±2 Further consideration of Nb-18 is given in Section 6.1.

In Figure 3, Org. L's measurements of the third batch, 1 mg strip with 19.6 ppm Ta, show their greatest variability, with their own standard deviation of 5.2% and differences amongst them exceeding $2\sigma_{meas}$, albeit with all data less than 7.3% from the mean. As for the other participants, there is good agreement with all measurements within $\pm2\sigma_{pop}$ of the mean and only three more than σ_{pop} from the mean. It is noted that despite the potentially discrepant mass measurement by Org. I for Nb-23 (Table 1), the specific activity obtained is in good agreement with the rest of the batch.

The fourth batch, shown in Figure 4, is the first of two sets of 8 mg disk 0.3 ppm Ta specimens. Being the smallest batch, it contains four specimens, of which only three were measured by two participants, and it has the largest σ_{pop} value of 7.9%. Compared to a measurement uncertainty of ±2.6%, data from Org. L for each of these samples are mutually consistent suggesting no sample-to-sample variability. For the other participants, this batch . It is appears to provide the most polarised inter-comparison with data split between two high values from Orgs. D and E (Nb-30 and Nb-31), of which the former has a significantly larger σ_{meas} of ±11.5%, and a markedly lower value from Org. G (Nb-28) with modest σ_{meas} noted, however, that with its relatively large uncertainty, Org D's measurement is nonetheless consistent with the lower values. Finally, Org. E's mass measurement for Nb-31 is 10% lower than the corresponding Org. L value suggesting that had Org. E obtained the same mass measurement, the agreement would have been much better although still high.

For the fifth batch (8mg disk 0.3 ppm Ta), Figure 5, the absence of sample-to-sample variation is demonstrated by Org. L's measurements, which show mutual consistency within $\pm\sigma_{pop}$ of 2.2% compared to measurement uncertainties of ±2.6%. Overall, the batch shows excellent agreement over eight samples, fourteen individual measurements and four organisations (Orgs. A, C, K and L). The level of agreement is represented by the lowest σ_{pop} (2.2%) and the fact that all but three of the individual measurements lie between the mean and $\pm\sigma_{pop}$. Unlike the first four batches, this batch contained three pairs of samples measured by the same organisations, namely: Nb-41 and Nb-42 (Org. K), Nb-45 and Nb-46 (Org. C) and Nb-51 and Nb-52 (Org. A). It is therefore interesting to note that in each pair, variations seen in the corresponding Org. L data appear to be reproduced, including the modest shape apparent at either end of the data.

Figure 6 shows the specific activity measurements for the sixth batch (8mg disk 4 ppm Ta samples). The largest of the six batches with ten samples plotted, the data from Org. L are all consistent within their measurement uncertainties of ±2.6%, confirming the absence of sample-to-sample variability. There is excellent agreement between the fifteen measurements provided by Orgs. A, C, G and L, all of which agree within ±1.4% of 8.08×10^{6} Bq g^{-1}. However, results for Nb-35 and Nb-40 (Orgs. E and D, respectively) are more discrepant, contributing to higher mean for the batch as a whole (8.30×10^{6} Bq g^{-1}) and a significantly larger σ_{pop} of ±6.2%.

6.1 Potential Outliers

Discussion of the specific activity results in Section 6 has identified potentially outlying values, in particular Nb-18. To determine whether any of the data should be considered as outliers, statistical tests have been performed on the results of each of the batches using both Inter-Quartile Range and Grubbs test [3]. The results of both tests confirmed that only Org. E's result for Nb-18 should treated as an outlier, justifying its exclusion from the mean and σ_{pop} calculated for Batch 2.

Further investigation into Org' E's result for Nb-18 showed that had its specific activity been calculated with the mass value measured by Org. L, the specific activity measurement would reduce from 3.50×10^{7} Bq g^{-1} to 2.96×10^{7} Bq g^{-1}. At this value, it would be 2.7 σ_{pop} greater than the mean, but would not be indicated as an outlier in either of the tests applied.

6.2 Distribution of Discrepancies

Figure 7 presents histograms showing the distributions of measurement discrepancies relative to the mean for each batch expressed as multiples of σ_{meas} (to the left) and σ_{pop} (to the right). The data are presented so that the distributions are evident both for individual batches and the comparison as a whole.

Compared to measurement uncertainties, it is clear that the majority of data (34 out of 40) are clustered between $-3\sigma_{meas}$ and +1 σ_{meas}, suggesting that for these measurements, at least, the uncertainties allocated are representative of the measurement precision. However, the remainder of the data (including one not shown because it is out of range of the graph) appear significantly high creating a somewhat skewed distribution.

Unsurprisingly, the comparison of discrepancies relative to standard deviation, σ_{pop}, provides a narrower distribution for the obvious reason that σ_{pop} represents the distribution of all of the data. However, the comparison relative to σ_{pop} provides additional insight showing the distribution to be very peaked at $-1\sigma_{pop}$, with 23 out of the 40 values concentrated in the three intervals centred on this, with the rest of the population lying above. This suggests that:

- The asymmetric tendency to measure high is present in a substantial proportion of the data, and not just the more obviously discrepant values identified in the comparison of relative to σ_{meas};
- The asymmetry observed in the discrepancies has pulled the mean upwards, with the consequence that the modal values (represented by the peak at $-1\sigma_{pop}$) appear as under-measurement when, in fact, they are most probably accurate.

6.3 Normalised Results

Table 3 compares the relative performance of each organisation by normalising the value it obtained for each batch, S_i , relative to the mean of all measurements for that batch and expressing it as a deviation (in %), i.e. $[S_i - \langle S_i \rangle_{batch}] / \langle S_i \rangle_{batch}$. The mean batch values used in the denominator here are those in Table 2 and indicated by the solid horizontal lines plotted in Figures 1 to 6. In doing this, the averages obtained over all of the batches measured represent the overall performance of each organisation, effectively aggregating the observations made in the section above on Figures 1 to 6. In addition to this, σ_{org}, the standard deviation of each organisation's normalised results is given, to provide an indication of the variability of each organisation's performance from batch to batch. The average value of each organisation's measurement uncertainty, σ_{meas}, is also included to help gauge the significance of these results. The key observations are as follows:

- The majority (7 out of 11) of organisations' mean discrepancies are less than their σ_{meas}, supporting the uncertainties allocated. Of those that aren't, two are comparable and the other is Org. E the most discrepant organisation.
- Excluding the value of 10.3% for Org. E, values of σ_{org} are between 0.9% and 5.1%, lower than the range of σ_{pop} values, which lie between 2.2% and 7.4% (Table 2). This suggests that each individual organisations' results tend to be systematically removed from the consensus (to some extent) with some variation around that represented by σ_{org}. In contrast, the batch-to-batch variability is larger because systematic differences embedded within each organisation's data are not correlated from one to the other. This was also observed in [1].
- Org. E appears to systematically over-predict all of the samples it measured, by on average 13%. and significantly greater than the declared measurement uncertainties. This is evident as a clear outlier in Figure 2, as well as values markedly above the distribution in Figures 1, 3, 4 and 6.
- Org. D over-predicts both of the samples it measured by on average 10%, lying clearly above the distribution of data in Figures 4 and 6. However, this is mitigated by substantially larger measurement uncertainties.
- Mean normalised results for all other organisations are mutually reconcilable within $\pm 2\sigma_{meas}$.

To assess the impact of the two most discrepant organisations on the inter-comparison as a whole, the final row in Table 3 shows how the mean normalised deviation for each organisation is changed if Org. D and Org. E are excluded from the calculation of the mean value for each batch. The result is a tighter distribution amongst the remaining organisations with noticeably less indication that they have under-measured relative to the mean. Doing this does raise the possibility that Org. H has also over-measured, being the only one of the remaining participants to be more than σ_{meas} from the mean.

6.4 Sensitivity to Measurement Methods

Table 3 also records the method used by each participant showing that a subset of seven used sample dissolution techniques, "D" and a subset of four used solid sample techniques, "S". For each subset, it provides the average of the normalised specific activity deviations (i.e. $[S_i - \langle S_i \rangle_{batch}] / \langle S_i \rangle_{batch}$) for each batch as well as the data as a whole.

t-Test results for each batch, and the data as a whole, indicate the likelihood of the subsets of data from the two methods having the same mean. Although Batches 1 and 2 appear to show differences between mean values of the the two subsets greater than 5%, the t-Test results suggest little to no difference evident between any of the individual batch results, and none for the data overall. It is important to note that the distribution of participants between batches is not consistent (e.g. Batches 4, 5 and 6 were measured by fewer participants), which this has the potential to influence the outcome of this comparison from batch to batch.

Whilst limitations in the dataset are acknowledged, the inference reached here is that there is no evidence to suggest systematic differences between organisations on the basis of the techniques deployed.

6.5 Tantalum Concentration

One of the objectives of the inter-comparison was to provide batches of specimens with a range of tantalum concentrations to determine whether this had any bearing on the ability of participants to provide consistent results, whatever the counting technique used. Despite the limitation on the distribution of batches between the participants identified above, it would be expected that any adverse effects of increasing tantalum would be evident as increased variability in the results and/or higher numbers of outliers. On inspection, the data presented in Table 2 show no evidence of any increase in σ_{pop} associated with increasing tantalum concentration, and a qualitative review of Figures 1 to 6 shows no greater tendency for there to be more outlying values in the batches with highest concentrations. These observations were supported by simple regression of σ_{pop} and batch range (i.e. highest minus lowest) vs. tantalum concentration, which indicated no significant correlation.

Therefore, whilst recognising its limitations, this inter-comparison exercise has produced no evidence that tantalum concentrations up to c. 20 ppm have affected the accuracy with which participants have measured the specific activity of ^{93m}Nb.

Table 3 Normalised Specific Activity Measurements

	Org L	Org K	Org J	Org I	Org H	Org G	Org F
Technique	S	D	D	D	S	S	D
Batch 1	-6.2%	-	-6.9%	0.3%	1.9%	-4.9%	2.7%
Batch 2	-3.0%	-	-3.2%	2.3%	5.0%	-4.4%	3.3%
Batch 3	-1.5%	-	-8.2%	-4.6%	11.8%	-5.0%	-1.1%
Batch 4	-7.3%	-	-	-	-	-6.3%	-
Batch 5	-1.1%	1.0%	-	-	-	-	-
Batch 6	-2.6%	-	-	-	-	-4.0%	-
Mean.	-3.6%	1.0%	-6.1%	-0.7%	6.2%	-4.9%	1.6%
σ_{org} (2)	±2.6%	n/a	±2.6%	±3.5%	±5.1%	±0.9%	±2.4%
Av. σ_{meas}	±2.6%	±5.1%	±7.0%	±3.0%	±4.1%	±3.5%	±3.9%
Rev. Mean(1)	-1.5%	1.0%	-4.9%	0.5%	7.6%	-2.3%	2.9%

	Org E	Org D	Org C	Org A	Mean D	Mean S	t-Test prob
Technique	D	S	D	D			
Batch 1	13.2%	-	-	-	2.3%	-3.1%	32%
Batch 2	30.8%	-	-	-	8.3%	-0.8%	33%
Batch 3	8.6%	-	-	-	-1.3%	1.8%	65%
Batch 4	6.2%	7.4%	-	-	6.2%	-2.1%	n/a
Batch 5	-	-	-1.7%	1.3%	0.2%	-1.1%	n/a
Batch 6	6.4%	12.7%	-4.5%	-1.8%	0.1%	2.0%	77%
Mean.	13.0%	10.1%	-3.1%	-0.2%	0.8%	1.9%	72%
σ_{org} (2)	±10.3%	±3.7%	±2.0%	±2.1%			
Av. σ_{meas}	±1.1%	±11.5%	±4.0%	±2.8%	-	-	-
Rev. Mean(1)	16.0%	15.5%	-1.6%	1.4%	2.2%	4.8%	64%

Note: (1) Results normalised vs. batch average values excluding Org. D and Org. E.
(2) σ_{org} is the standard deviation of normalised specific activity measurements for each organisation.
(3) Where t-Test results are not given, there were insufficient data in one of the subsets.

7 Suggestions for Improvement

As the first international Round Robin inter-comparison of the Dosimetry Community's facilities for the measurement of the activation product ^{93m}Nb, this exercise has provided valuable data and insights. Because of the relatively late adoption of this radionuclide by the Reactor Dosimetry Community, and because of the challenges associated with performing multi-participant measurements on samples that have traditionally required destructive techniques, it is unsurprising that such an activity has not been undertaken previously.

Being first in kind, it is also not surprising that, with the benefit of hindsight, there are ways that this exercise could be improved in any future inter-comparison. The points below provide a non-exhaustive list of the more important lessons learned:

- Although the technique of irradiating batches of near identical samples was successful, the batch technique would have been much more effective if each participant had received and measured an equivalent set of samples, each drawn from the same of batches. The inconsistent distribution of samples between the participants has weakened some of the conclusions of the exercise.
- Increasing the numbers of dosimeters received and measured by each participant would have improved the statistical significance of the results.
- Recognising that about a third of participants used non-destructive (solid) techniques, the sharing of samples between those organisations could have been possible before providing to an organisation using

dissolution techniques. Had this been realised, it would have been possible to make the limited number of samples go further between the participants and may have made the conclusions more compelling.

- Whilst the provision of samples with different geometries has provided important variations between the measurements performed, a more systematic investigation of the impacts of geometry would require careful design. In particular, future inter-comparisons would benefit from a greater range of specific activation values, including some that would stretch the abilities of the facilities.
- A more conclusive investigation into the effects of tantalum concentrations would have been obtained if there had been a better distribution of tantalum values between the batches, with more at higher concentrations in particular.
- Damaged samples should not be forward to participants, if it can be avoided. To that end, an increase in the number of samples irradiated would provide some mitigation against the consequences of damaged specimens (dosimeters broken during their recovery from irradiation could have been substituted).
- A protocol for dealing with the most outlying measurements should be agreed in advance.

8 Conclusions

Following on from an initial Round Robin inter-comparison of gamma spectrometry measurements reported in 2014, the European Working Group on Reactor Dosimetry has undertaken a Round Robin exercise to compare measurements of the important activation product ^{93m}Nb reaction, which plays a key role in underwriting reactor dosimetry assessments.

The results presented anonymously in this paper show that:

- An inter-comparison of ^{93m}Nb specific activity measurements performed by 12 European Organisations has been successfully carried out, overcoming the challenge presented by the use of destructive (i.e. dissolution) techniques by the majority of participants.
- The technique of irradiating batches of near identical samples was successful. The use of a single organisation to provide a control against sample-to-sample variation was a key part of this.
- Comparisons performed on measurements from six batches of samples (each with its own geometry and tantalum concentration) show agreement to within standard deviations ranging from ±2.2% to ±7.9%.
- The majority of measurements are clustered within around 2% of the modal value for their batch.
- The distribution of measurements relative to their respective batch mean values is skewed, with few measurements below the modal value, and a tail comprising a small but significant minority of participants tending to over-measure.
- Although mass measurements were shown to be consistent for the majority of specimens, some of the most discrepant mass data were shown to be associated with discrepant specific activity measurements.
- From the data provided, there is no evidence to suggest that either measurement technique or tantalum concentration have had any impact on the accuracy of the measurements.

Finally, the inter-comparison described in this paper has provided valuable data and insights into the measurement of this key radionuclide. However, being first in kind, there are a number of ways that this exercise could be improved, seven of which are described in the paper.

References

1. D.A. Thornton, J. Wagemans, A.C. Fernandes, J.M. Girard, D.P. Kis, V. Klupák, R. Mutnuru, H. Philibert, G. Rousseau, J.P. Santos, T Serén, E. Zsolnay, *The EWGRD Round Robin Measurement Exercise*, submitted to the 15th International Symposium on Reactor Dosimetry, EPJ Web of Conferences **106,** 06005 (2016)
2. D.A. Thornton , C. Domergue A. Fernandes P. Frajtag, P. Haroková, D.P. Kis, V. Klupák, V. Kochkin M. Lépy, M. Mitev, R. Prokopowicz, V. Radulović, B. Straathof, J. Wagemans, E. Zsolnay, The Second EWGRD Round Robin: Inter-Comparison of Gamma Spectrometry Measurements on Activation Dosimeters, submitted to the 17th international Symposium on Reactor Dosimetry, Lausanne, 2023
3. F.E Grubbs, Procedures for Detecting Outlying Observations in Samples, Technometrics, Vol. 11, No. 1, February 1969

Acknowledgements

The EWGRD would like to thank the participating authors and their organisations for their support in the completion of the exercise and the production of this paper. In particular, the authors would like to acknowledge the support of CEA, NCBJ and SCK-CEN whose commitment and initiative made the exercise possible.

AUTHOR INDEX

EDP Sciences – Web of Conferences
17, Avenue du Hoggar
Parc d'Activité de Courtabouf
BP 112
F-91944 Les Ulis Cedex A
France

ISBN 978-1-7138-7052-4

9 781713 870524

CURRAN ASSOCIATES INC.
proceedings.com